Science, Rationality, and Neoclassical Economics

Science, Rationality, and Neoclassical Economics

L. D. Keita

Newark: University of Delaware Press
London and Toronto: Associated University Presses

© 1992 by Associated University Presses, Inc.

All rights reserved. Authorization to photocopy items for internal or personal use, or the internal or personal use of specific clients, is granted by the copyright owner, provided that a base fee of $10.00, plus eight cents per page, per copy is paid directly to the Copyright Clearance Center, 27 Congress Street, Salem, Massachusetts 01970. [0-87413-410-2/92 $10.00 + 8¢ pp, pc.]

Associated University Presses
440 Forsgate Drive
Cranbury, NJ 08512

Associated University Presses
25 Sicilian Avenue
London WC1A 2QH, England

Associated University Presses
P.O. Box 39, Clarkson Pstl. Stn.
Mississauga, Ontario,
L5J 3X9 Canada

The paper used in this publication meets the requirements
of the American National Standard for Permanence of Paper
for Printed Library Materials Z39.48-1984.

To T. K. and L. K.

Library of Congress Cataloging-in-Publication Data

Keita, L. D.
   Science, rationality, and neoclassical economics / L. D. Keita.
     p.   cm.
   Includes bibliographical references and index.
   ISBN 0-87413-410-2 (alk. paper)
   1. Neoclassical school of economics.   2. Decision-making.
3. Rationalism.   4. Knowledge, Theory of.   5. Science—Methodology.
I. Title.
HB98.2.K45   1992
330.15'7—dc20   90-50455
                                                                             CIP

PRINTED IN THE UNITED STATES OF AMERICA

Contents

Preface	6
1. Introduction	9
2. Knowledge and the Theory of Science	13
3. The Structure and Proof of Scientific Theories	27
4. The Classical and Neoclassical Methodology	41
5. Ordinal Utility Theory and Contemporary Neoclassical Economics	57
6. General Equilibrium Theory—An Analysis	83
7. The Postulate of Rationality and Neoclassical Economic Theory	94
8. "Positive" Neoclassical Welfare Economics	113
9. Alternative Methodologies	122
10. A Theory of Optimal Decision Making	133
Notes	152
Bibliography	169
Index	177

Preface

There is no doubt that economics is an important social science, given its capacity to analyze at base humankind's relationship to the resources of the environment. It is this human response to the resources of the environment, in whatever contingent circumstances, that shapes civilisations and cultures, and serves as the leitmotif of history. In fact any human activity whether in the form of technology or the arts could be analyzed from an economic standpoint. Some theorists have even argued that the content of philosophical ideas derives from the economic conditions of life.

Given the importance of economics to human social life I see some virtue in analyzing the neoclassical economic theory, one of the dominant economic systems of our time. Its advocates claim that this economic theory is scientific or at least on the way toward becoming a science. The thrust of my analysis is that the claims to scientific status on the part of neoclassical theory could be questioned epistemologically. The bulk of this text is concerned with this epistemological analysis.

I have shared the ideas expressed in this text with colleagues, especially K. K. Dompere, H. Harriot, and A. Klamer. Their comments, together with those of unknown referees, were helpful. I must also thank, among others, C. P. Foray, Principal of Fourah Bay College, who granted me leave from regular academic duties to complete this text; also S. Lithgow for efficiently typing this manuscript.

Science, Rationality, and Neoclassical Economics

1
Introduction

The question of whether there can be a genuine science of human behavior is a perennially intriguing one. The evident success of research in the physical and biological sciences has encouraged the argument that the methodology of research in those areas could be successfully employed in the analysis of human behavior. The idea of this thesis was first proposed by the empiricist philosophers of the past and the positivists of more recent times. The basis for this unity of science approach is that any empirically observable phenomenon is subject to scientific analysis according to a general methodology of research.

Yet because of the limitations in scientific research on the observable behavior of human beings, some theorists have argued for a qualitatively different methodology of research for human behavior. Herein lies the genesis of the *Geisteswissenschaften*, hermeneutic, and the more recent interpretive research schools that make the case for special techniques of analysis more appropriate for the analysis and understanding of human behavior. The view here is that human behavior, though objectively given, can be fully understood only in its subjective dimensions.

But the neat formalism of the empirical sciences continues to influence analysts of human behavior. Thus, for example, the ability of neoclassical economics theorists to formulate impressive mathematical models of decision making is viewed as hope for those who expect the sciences of human behavior to vindicate the unity of science thesis. In fact, the neoclassical research paradigm is dominant among professional economists, and despite intellectual challenges from a number of quarters does not appear seriously challenged in terms of adherents at the major centers of research. It is generally held by its supporters that economic behavior can and must be studied scientifically. It is also claimed that scientific economics—that is, neoclassical economics—engages in empirical and objective analysis of economic behavior. In this regard, scientific economics must be distinguished from normative economics, which is usually viewed as embracing welfare economics.

Yet, given the fact that neoclassical economic theory has not yielded the expected results in terms of prediction and explanation—the two most important requirements for a scientific theory—an increasing number of critical theses and alternative research paradigms are being developed.[1] It should be noted, though, that most critiques of neoclassical economic theory emphasize the fact that its predictive and explanatory record has not been particularly encouraging. But there have been very few studies that examine the theoretical foundations of the theory to determine whether the discipline needs a complete restructuring or whether the research venture of attempting to formulate a science of economic decision making is indeed possible.

The result of this epistemological myopia is that the alternative research paradigms such as the Institutionalist, neo-Austrian, behavioral, and Marxian commit theoretical errors similar to those of the neoclassical research theorists themselves. The approach I undertake in this study is nonorthodox in the sense that not only do I attempt to argue that the neoclassical economics claim to scientific status cannot be supported, but also seek to show why this is the case from a theoretical and structural point of view.

Specifically, I attempt to show that the key postulate of rationality from which the axiomatic structure of neoclassical economics is derived cannot be viewed as a scientific postulate. It renders the axioms and derived theorems of neoclassical theory essentially normative. I also argue that any systematic study of economic behavior is necessarily value laden. I show, too, that in spite of the fact that human decision making appears to be a proper candidate for scientific analysis, the methodology of investigation adopted by the neoclassical economics theorist has not been successful. The substitution of the postulate of rationality as a formalizing principle for measurable, empirically derived propositions has proven to be a costly heuristic for a discipline that aims to be scientific.

Perhaps the solution to the problems of economic theorizing lies in accepting the idea that any science of human behavior must be necessarily prescriptive in the sense that the theorist is constrained to make certain evaluative assumptions to interpret or explain observed behavior. Recall that a proper explanation of human behavior must appeal to the ideas of motive and purpose, which are yet to be empirically quantifiable.[2] And given the lack of the empirical inaccessibility of motives and purposes, the theorist is forced to imagine just what they might be. Yet he or she can construct only one explanatory theory at a time. However, no such single explanatory theory can really suffice to explain and predict the great multiplicity of human choices. It is here that the theoretical usefulness of the purportedly predictive but evidently normative heuristic of the rational decision maker is evident.

I believe that we are forced to recognize that human behavior, though goal directed, is not essentially predictable. The complexity of the neuronic structure of the human brain, which is at the source of the subjective notion of choice, greatly complicates the situation. The point I make is reinforced by the fact that no theory of decision making founded on principles of choice has been shown to be predictively (hence explanatorily) unproblematic when applied to cases of actual choice making. The decision theorist may, of course, not make references to the idea of rationality in theory construction, but this would not belie the fact that any theory of decision making must be founded on a specific set of choice rules. It could not be otherwise. The belief held by some theorists that the problem of decision-making theory could be solved if it were possible to devise some model of decision making that would maximally conform to actual choices is erroneous. As long as human beings perceive themselves as being capable of choosing among alternatives whenever confronted with any decision schedule, then any model of decision making is necessarily normative. In this regard, variants of rational decision making as exemplified by the various interpretations of expected utility theory,[3] though useful for their added insights into the psychological vagaries of human choice making, cannot hope to salvage the claims of neoclassical economic theory to scientific status.

As suggested above, the best solution to the problem of formulating sciences of human behavior in general, or a science of economic decision making in particular, would be to accept the notion that only normative models of optimal decision making could be sufficiently developed to serve in any meaningful scientific capability. This would mean that theorists of economic behavior are necessarily committed to some particular set of normative assumptions. The "science" of economics should be best viewed, therefore, not as "the rational allocation of scarce resources" but as "the rational allocation of scarce resources according to a prescribed set of value assumptions." One could paraphrase this definition to mean that economics is the study of human economic welfare.[4] The theoretical distinction between positive and normative economics is thus seen to be questionable. The proper study of economics should, therefore, make reservations for political and sociological variables, normally ignored by theorists of neoclassical economics. More importantly, however, the aim of an adequate economic theory would be one that offers a normative guide to optimal economic decision making within the constraints of the social collectivity. The economist, therefore, should not regard himself or herself as a purely empirical scientist but rather as an applied scientist much like the physician whose goal is to optimize patients' physical health given the constraints of their actual health conditions and the scientifically derived curative techniques available. Were the physician to accept a research paradigm similar to that of the neoclassical economist, his or her avowed aim would be to diagnose patients without any explicit reference to their actual health status but according to some ideal model, which (though prescriptive) would be viewed as objective and scientific. But the practice of medicine would be reduced to a travesty were this mode of operation the actual case. Thus, in order to restore meaningfulness to economics as a research discipline, the economist must begin his or her research with the key value notions of welfare, equity, and so on very much in his or her mind.[5] The goal of economic theory should be then to formulate the conditions for the optimizing of human capital and human welfare for each individual economic agent.

The program I choose to follow in the chapters ahead are as follows. I shall first engage in a discussion of the methodology of scientific research. This discussion will serve as a kind of template against which neoclassical economics' claim to scientific status will be judged. I shall then discuss the efforts made by students of human behavior to establish a science of human decision making in the course of history. In the context of economics, the shaping role of precontemporary theorists and cardinal utility advocates will be discussed. Next, I shall examine the central role of ordinal utility theory and the postulate of rationality in the development of modern neoclassical economics. Finally, discussions of the general equilibrium and positive welfare economic theories followed by a statement on the current theories rival to neoclassical economics will be pursued before I propose an alternative approach that could be a possible solution to the problem of formulating an adequate theory of economic decision making.

2
Knowledge and the Theory of Science

Scientific Methodology: A Historical Overview

It is almost a truism to state that human beings historically have equated knowledge about empirical phenomena with true claims about such phenomena. But the problem concerning the acquisition of knowledge is that the obtaining of facts about phenomena requires more than just immediate sensory observation. Knowledge about phenomena requires not only an appeal to the idea of explanation but also to that of prediction. Yet, although explanation and prediction appeal to those aspects of phenomena which are not directly observable, it is these very aspects that are the basis for the validation of theories about directly observed phenomena.

It is on the basis of sensory evidence that theoreticians of knowledge have seen fit to distinguish confirmable empirical propositions from those which were metaphysical, mathematical, or ethical. This is not to deny, of course, that some thinkers have sought historically to argue that genuine knowledge was derivable not from sensory observations but from the nonempirical claims of mathematical, ethical, or metaphysical propositions. Students of the history of ideas would, no doubt, recall the Platonic theory of forms, which sought to ground genuine knowledge in nonempirical abstract entities. One might note, too, the special place ascribed by premodern epistemologists to mathematical propositions and some logically defensible ethical postulates.

Finally, however, empirical knowledge, though disparaged by some premodern theorists, gained in ascendancy because of its pragmatic yield. Empirical knowledge validated and buttressed by quantitative reasoning proved decisive in the development of technology. In turn, it was an improved technology that yielded the possibility of employing novel experimental techniques the results of which aided in its further development. It is instructive to note that the critical transitional period occurred at the inception of the Renaissance. Theories in ethics and religion were increasingly relegated to qualitatively different areas of investigation, while novel methodologies of empirical inquiry were being formulated by such theorists as Galileo, Descartes, and Newton.

It is in this historical context that one recognizes the novel paradigms of research introduced in Newton's *Principia* and Descartes's *Discourse on Method*. The epistemological unity of research promoted by the assumption of a blurred distinction between animate and inanimate phenomena was questioned and rejected by all those who supported the methodological independence of research into inanimate phenomena. Traditional explanatory notions such as "vis viva," "sympathy," and so on were rejected in favor of more testable concepts.

The same thinkers were also quite clear in establishing lines of demarcation between the required methodology of investigation of the natural world and other subjects of inquiry such as religion and ethics. In fact, a distinction was even made between principles of ethical behavior and inquiries into possible explanations of the human "passions."[1] Newton and others were concerned to purge the new science of mechanics of any appeal to occult principles, notwithstanding the former's interest in astrology and alchemy. In fact, there was even skepticism expressed against Newton's theory of gravitation given a perceived resemblance between it and other scholastic concepts such as sympathy and antipathy.

Newton's theory of gravitation was vindicated only when he pointed out in his theory of gravitation that notions such as "attraction" were not to be understood in a physical causative sense but only in a mathematical sense. Furthermore, Newton was careful to demonstrate that the methodology of the new mechanics did not seek to inquire into the ultimate causes of phenomena or to explain or describe phenomena beyond what was empirically observable. One recalls, in this connection, his well-known phrase "*hypotheses non fingo.*" As he put it:

> But hitherto I have not been able to discover the cause of those properties of gravity from phenomena, and I frame no hypotheses; for whatever is not deduced from the phenomena is to be called an hypothesis; and hypotheses, whether metaphysical or physical, whether of occult qualities or mechanical, have no place in experimental philosophy. In this philosophy particular propositions are inferred from the phenomena, and afterwards rendered general by induction.[2]

Thus, what was characteristic of the postscholastic research paradigms was the emphasis placed on the formulation of explanatory theories of observable phenomena based on sound experimental procedures. No hypothesis based on speculation was to be considered acceptable as scientific unless confirmed by experiment. The search for the ultimate causes of phenomena was also to be disregarded on the grounds that proof of the existence of such could not be forthcoming. Needless to say, evaluative hypotheses concerning human conduct had no place in the new mechanistic world view for they belonged rather to theology, the province of religious belief.

It is useful to point out the roles of such methodologists as Bacon, Galileo, and Descartes in the formulation of the novel paradigm. Bacon, of course, stressed the importance of empirical observation for the formulation of hypotheses about phenomena, whereas Descartes recognized the necessary role of mathematical expression and deductive reasoning in any proper scientific study. These two methodological emphases were no doubt reflected in the two dominant philosophical systems of the period: rationalism and empiricism. Actual scientific practice eventually came to represent a fusion of both theoretical perspectives. Contemporary scientific practice, no doubt, recognizes not only the necessity of the empirical basis for the formulation of scientific hypotheses but also the deductive certitude afforded by the expression of such hypotheses and their implications in the quantitative language of mathematics. Furthermore, the use of mathematics (the archetypical rationalist discipline) allowed the researcher to demonstrate the evident superiority of quantitative over qualitative expression.

It is instructive to note, though, that even after a distinct qualitative break with pre-Renaissance methodologies of investigation, the new methodology of investigation nevertheless was subject to two different emphases ever since its near unanimous adoption by the theorists and practitioners of scientific investigation. The two methods were the method of hypothesis and the method of inductive inference. A similar epistemological divergence may be noted today between the instrumentalist and realist approaches to scientific investigation. Larry Laudan argues that the method of hypothesis (the hypothetico-deductive method in contemporary parlance), though initially popular in the seventeenth century, fell into disfavor in the eighteenth century only to be rehabilitated once more in the nineteeth century.[3] I am inclined to believe, however, that there may be some exaggeration here. In experimental matters concerning mechanics, the method of inductive inference was the more preferred, but this was quite inadequate for analyses entailing the recondite causes of phenomena. Granted the inadequacies of the instrumentation of the times, most research that appealed to explanations that could not be directly confirmed—that is, microscopic or subatomic phenomena—necessarily relied on the method of hypothesis for theory formulation.

Of course, the history of science can make reference not only to strict methodologists of science but also to those scientists themselves who chose to write on methodology. Some good examples of this are Newton, Mach, and Einstein. However, what is evident from the writings of these important figures is that scientific theories should necessarily be empirically grounded in the two areas of prediction and explanation. And theories that appealed to explanatory concepts that could not be empirically confirmed were to be treated as plausible conjectures until sufficient evidence was forthcoming. Practitioners in science who have also written seriously on the issue of methodology have generally regarded some theory of realism as the preferred epistemological outlook. Instrumentalism is regarded only as a second best heuristic for the formulation of theories until such theories are reinforced or discarded on account of convincing empirical evidence. Thus, the key concepts in the scientific enterprise since the beginnings of the modern era are empirical evidence, quantification of evidence, and testability of hypothesis under controlled conditions. The implications of these requirements are that theories founded on metaphysical or evaluative assumptions are not generally regarded as acceptable.

Given the above discussion, it is useful now to discuss the models of investigation that contemporary theorists of science have formulated to express the general approach to scientific investigation in the modern era. In a strictly formal way, two quite simple models of scientific theory construction have been developed in recent times to explicate the practice of scientific research: the deductive-nomological (D-N) and the inductive-statistical (I-S) models of explanation.[4] The D-N model states that any genuine scientific explanation can be formulated in terms of a logical relationship between a theory's explanans (theory and experimental design) and the event to be explained (explanandum). According to this purely schematic model, the most important component of the explanans is the set of general laws that are central to the explanation of the explanandum.

According to the D-N model, the explanandum is logically implied by the explanans, thereby justifying its predictability. Furthermore, the explanandum is accounted for in terms of its explanans, that is, its relevant theory and experimental design. Similarly in the case of the I-S model the explanandum is related to the explanans according to some degree of inductive probability.

The obvious simplicity of the D-N and I-S models have led to much useful criticism. Some authors have argued that not all explanations that conform to the D-N model may be regarded as genuine explanations,[5] and that the assumed symmetry between explanation and prediction do not apply to all areas of scientific investigation.[6] I should not want, at this point, to engage in a detailed defense of the D-N and I-S models; I state briefly instead that the criticisms raised against the idea of the D-N explanation especially are not defensible. Quite clearly, not every explanation should be regarded as a scientific explanation nor should every account of some event entailing deductive inference be categorized as scientific. Explanation, deductive inference, prediction, and so on are to be regarded as necessary criteria that any scientific theory must uphold, but they are not sufficient. A necessary concomitant to any genuine scientific explanation is that the explanandum to be tested must be confined to the controlled conditions of the experimental laboratory. It is only within the confines of the controlled experiment that a theory may be tested repeatedly in all its variations. In this connection, I should want to argue that much of the criticism leveled against the D-N model is based on the assumption that certain research areas, which I hold to be not purely scientific in the sense of the criteria stated above, are assumed as such. Examples are the social sciences and such areas of research as astronomy and archaeology. But, I should want to argue, too, that given the constraints on controlled repeatable experiments on human beings, the social sciences, in general, should be best viewed as quasi-scientific rather than fully scientific. Similarly, areas such as astronomy and archaeology ought to be regarded as applied sciences than otherwise. There are no strictly manipulable conditions possible for astronomy and archaeology; what we have is a heavy reliance on theories of the physical sciences.

Similar kinds of arguments could be made against the inductive-statistical (I-S) model on the grounds that the uncontrolled conditions in which statistical explanations are offered do not satisfy necessary criteria for scientific status. In fact, I argue that the I-S model and other similar models cannot be viewed as representative of scientific theories since such models properly support only partial explanations rather than complete explanations as in the case of the laboratory sciences. The relationship between the premises of the I-S model—that is, its explanans—and its explanandum is determined by some probability relationship r. This probability relationship (usually less than 1), guarantees that the model's premises need not be rejected if the explanandum eventuates or not. But surely, if this is permissible, then the model states at best statistical correlations rather than explanations. I am inclined to believe that probabilistic explanations are not complete explanations in a genuine scientific way.

Consider the following example: The probability that individual X would contract lung cancer if he or she satisfies criteria $a_1, a_2, \ldots, a_n$ is .95. X satisfied criteria $a_1, a_2 \ldots, a_n$. We infer, therefore, that the probability that X would contract lung cancer is .95. Yet, what we have established here is not an explanation but merely a statistical correlation. As long as there exists some individual Y who satisfies criteria $a_1, a_2, \ldots, a_n$ but does not contract lung cancer, then the relationship between the explanandum, X contracted lung cancer, or Y did not contract lung cancer, and its relevant explanans is merely probabilistic not explanatory. It seems to me that a genuine scientific explanation is necessarily based on a confirmable causal relationship between some given event and its antecedents.

Wesley Salmon, who has done substantial research on the problem of statistical explanation, has recognized that there are problems with the I-S model of explanation. He rejects the view that I-S type explanations are acceptable on the grounds that they are potential D-N explanations.[7] Given the fundamental role of statistical laws in modern scientific theory, Salmon argues that a genuine defense of I-S type explanations is required. But it should be noted that whatever heuristic functions probabilistic or statistical type explanations serve, their epistemological acceptability is ultimately dependent on the confirmable empirical content of their claims. Salmon's defense of statistical explanations is firmly committed to this point of view.

It is useful to point out that the recent literature in the theory of science is no longer exclusively concerned with the establishing of formal models but also with the actual practice of science in the course of history. Note, too, that although as a rule historians of science have been minimally concerned with the theory or methodology of science, Kuhn's *Structure of Scientific Revolutions* initiated a new research trend that sought to integrate historical analysis with methodological investigations. In the course of this new trend, the ideas expressed by Feyerabend, Popper, and Lakatos became well known. The significance of the new approaches to the theory of science is that it affords evidence of the ways in which research standards have been established and modified in time. The more recent research in the theory of science also points out how the actual practices of past scientists have measured up to the current prescriptive criteria for scientific research.

The strongest advocate of the view that there has always been and ought to be diversity in the formulation of methodological standards has been that of Paul Feyerabend.[8] Kuhn, of course, emphasizes the historical path of science, whereas Popper argues on the other hand that the scientific enterprise ought to be judged essentially in terms of a strict methodology founded on the idea of falsifiability. Lakatos attempts to strike a balance between these two extreme views by arguing that although constraints on human epistemology may lead to a distinctly nonlinear historical path of science, it is extremely important to maintain methodological standards of research. These views are expressed in terms of Lakatos's idea of the methodology of scientific research programs. I am inclined to believe, however, that regardless of the facts about the actual path of science in history, theoreticians of science and scientists themselves always have sought to justify their claims about the world in terms of some set of criteria that set the greatest store by empirical evidence and confirmatory warrant. In fact, without a set of standards stressing empirical confirmation, one would be at a loss to explain the cumulative growth of scientific knowledge. Yet it must be pointed out, though, that the critical period that marked the formulation of methodological criteria employed by modern science is that of the post-Renaissance.

Thus, despite the fact that the research methodologies employed by contemporary scientists may not be exactly those of the time of Newton, these methodologies are indeed implied by such. The fact that some theories have been eventually rejected despite being accepted for lengthy periods of time by the scientific community can be attributed to the erstwhile limitations of experimental technology and the consequent respectability of instrumentalism as a research paradigm.[9] Implicit in the claims I make is the assumption that questions about methodology are not as intractable as some theorists are inclined to believe. As argued above, scientific theories are ultimately judged in terms of their confirmatory potential and empirical warrant. In this connection, I perceive no tension between the normative study of scientific methodology and historical analyses of the path of science in history. The cumulative growth of scientific knowledge affords adequate evidence that the history of science cannot be written without making references to the normative issue of methodology.

In order to evaluate properly the scientific claims of theories in the social sciences, it is necessary to formulate a set of methodological criteria against which social science in general and economics in particular will be evaluated. Given the very evident success of research in the natural sciences, researchers in the social and biological sciences have always sought to apply the methodology of the former to the latter's areas of inquiry. And in the light of the experimental constraints placed on the researcher in the social sciences on account of the special nature of the object under inquiry, some methodologists of social-science inquiry have argued that a qualitatively different set of research criteria are needed for inquiry in that area. Herein lies the basis for the historical divide between the positivists and those who embrace the phenomenological approach, that is, between those who argue that scientific research on human behavior should be qualitatively similar to research in the natural sciences and those who argue for a different methodology.

On the basis of the assumption that the intent of scientific research is to understand and interpret phenomena of the empirical world, it is understandable why many methodologists of research have sought to argue for one set of research criteria for both natural and social science. For it is the researchers in the natural sciences who have been able to establish the appropriate experimental criteria for the understanding and interpretation of empirical phenomena. Thus it should always be borne in mind that what is essential and necessary for scientific research and that which constitutes the basis for its success is its capacity for hypothesis testing. In other words, no theory or hypothesis may be viewed as being truly scientific unless there are experimental procedures that could either potentially confirm or question it.

Recent Methodologies of Scientific Investigation

Kuhn and Feyerabend

The discussion on the proof of scientific theories would not be complete unless some further reference is made to contemporary trends in the methodology of scientific research. Until the time of Kuhn, the dominant thesis was that science distinguished itself from other fields of inquiry in that it made progress cumulatively and was the best intellectual exemplar of the rational analysis of phenomena. Kuhn's important *Structure of Scientific Revolutions* had the effect of challenging the orthodox thesis of scientific progress with its claims that progress in science was not linear but determined by conflict between incommensurable paradigms. The result of this claim for relativism in scientific research was to raise theoretical questions about the nature of scientific proof. According to the Kuhnian thesis, such concepts as "empirical proof," "theory confirmation," and so on were not subject to objective valuation; they were determinable only from the vantage point of the paradigm.

Kuhn argues that progress in science is effected when a novel paradigm first becomes entrenched in time then is successfully challenged by a new theory (among others) when new anomalous data cannot be handled by the entrenched theory. This thesis, though interesting, leaves unanswered the most important question of scientific methodology: by what methodological mechanisms does science determine the validity of a theory in terms of its empirical content? For after all, the purpose of scientific research is to formulate theories that describe the world as accurately as possible, that is, to formulate theories with the greatest possible empirical content. Furthermore, Kuhn's thesis does not take into account the fact that the methodology of scientific research has improved greatly in time especially on account of more sophisticated instrumentation. It should also be stated that the theories to which Kuhn appeals to validate his thesis—as, for example, the Ptolemaic and phlogiston theories—should best be regarded as protoscientific rather than genuine scientific theories. In this regard then, Kuhn's so-called scientific revolutions should be properly described as making the transition phase between protoscientific and genuine scientific theories.

Of course, the literature on Kuhn's thesis is quite extensive, with other theorists gaining a reputation as they seek to refute or modify it. Feyerabend,[10] for example, has supported an extreme epistemological relativism in arguing that the history of science supports the prescriptive thesis that a proliferation of explanatory theories encourages the best scientific results. For Feyerabend, there should be no period in the phase of a theory's career that should be referred to as "normal science." Any sustained period of orthodoxy in research, according to this theorist, would tend to encourage dogmatism and the stifling of the creative spirit.

The essential point about the approaches of Kuhn and Feyerabend is that their theories of scientific proof are founded on the notion that the validity of a scientific theory is determined not by some set of objective rationalist standards but by sociological or even ideological considerations. Kuhn's key concept, it is recalled, is the research paradigm characterized by its own specificity and puzzle-solving activities. For Feyerabend, the guiding principle in research has been the subjective ideology of the members of particular research schools. This is the basis for the Kuhn-Feyerabend claim that novel theories in the history of science are often incommensurable with their predecessors. Thus, according to both theorists, Newtonian mechanics was incommensurable with relativistic mechanics. The result is that novel facts are often not capable of forcing the upholders of a given paradigm to effect a gestalt or conceptual switch, thereby leading to the embracing of the new theory. Since novel facts can always be reinterpreted to fit the old theory, new theories arise only when a new research paradigm attracts an increasing number of adherents on the grounds that it is perceived to solve puzzles more efficiently than its rivals. According to Kuhn, the supporters of increasingly challenged theories do not generally give up their old affiliations, they simply lose preeminence in their fields. Feyerabend, as epistemological Dadaist, would argue that supporters of theories that seem to be losing popularity should seek to increase their appeal by almost any means.

Now to return to the focus of this text, the question that must be asked is what relevance does the above discussion have on the methodology of the social sciences in general and economics in particular? A plausible response is that Kuhn's path-breaking text and the literature it generated focused more on the social sciences than on the natural sciences. I suspect that the reason for this is that, to a much greater extent than the natural sciences, the social sciences are riven by competing research paradigms. The problems associated with theory incommensurability, scientific objectivity, and so on are much more prevalent in the social sciences than in the methodologically more settled natural sciences. The interesting point about the response to Kuhn's text is that while the unity-of-science thesis has been traditionally espoused by the positivists, with the natural-science methodology as the ideal, the Kuhnian theory would argue that the cognitive claims of the theories in the natural sciences were epistemologically relative to their particular paradigms. Thus, it was in this climate of epistemological challenge to the positivist methodology that the neoclassical economist was forced to take stock of the theoretical assumptions of his or her discipline. But it was not only Kuhn who caught the attention of the natural and social scientist; the ideas of philosophers of science Karl Popper and his colleague, Imre Lakatos, on the methodology and history of science have been of much relevant interest also to the natural and social scientist concerned with questions of methodology.[11] I will now discuss their views on the methodology of science, with some comment on the research efforts of Laudan.

Popper and Lakatos

Karl Popper's prominence in the methodology of scientific research derives from his efforts to offer an alternative to inductive inference as the optimal epistemological instrument whereby scientific theories are validated. Popper's solution to the "problem of induction" is the argument that the aim of the theorist should not be to verify theories by means of inductive proof since such proofs are logically indefensible from the point of view that no logical proof could derive a valid universal statement from a finite set of statements confirming the event or phenomena described by the statement itself.[12] Popper argues that the key criterion that determines a genuine scientific theory is the falsifiability criterion. Thus, if a theory can specify clearly the conditions under which it could be falsified and if its experimental design clearly formulates the conditions for so doing, then the theory is a serious candidate for scientific status.

It should be noted that Popper's critique of inductive inference was concerned to offer an alternative to the then current logical positivist criterion of theory verifiability. According to the logical positivists, significant propositions were either analytic or synthetic. Analytic propositions were universally true purely according to their meanings, whereas synthetic propositions derived their truth content from their testable empirical content. But as Hume argued in his *Treatise of Human Nature*, claims about the empirical world could not be logically supported given the impossibility of affording a deductive proof of ampliative inference. Scientific claims are indeed claims about the empirical world and as such are subject to the skepticism generated on account of the logical problematic of inductive inference. The important epistemological question that issued from this situation was what could be the adequate bases for the determination and justification of scientific theories.

It is in this context that Popper's methodology of scientific investigation may be viewed as prescriptive in contrast with Kuhn's analyses, which were founded on a particular version of the actual history of science. Popper's program for the path of scientific research is that the goal of science should be to establish confirmable facts about the empirical world by the formulation of increasingly bold conjectures with high degrees of improbability. Such theories, according to Popper, have greater empirical content than theories with greater probability (i.e., lesser improbability). The status of such theories is to be determined by their capacity to withstand falsification through testing. And the hallmark of a good scientific theory is increasing empirical corroboration with respect to an ongoing battery of tests.

It is useful to point out that the main motivation for Popper's approach to the methodology of science was partially ideological: Popper was concerned to show that Marxist and psychoanalytical theories were not scientific, since such theories were impervious to potential falsification. He argues that whereas Marxism was once scientific but became nonscientific when it adopted "an immunizing strategy," psychoanalysis was nonscientific from its inception.[13] In the case of Marxism, Popper argues that Marxism as a scientific theory was falsified when predictions of the theory were not borne out in fact. And the persistent appeal to *ad hoc* hypotheses by Marxian theorists to salvage their theory is viewed by Popper as unscientific.[14]

While Popper's contribution to the methodology of science was principally prescriptive, Lakatos sought to apply Popper's falsificationist program to the actual path of scientific research in history. The result was the formulation of what is generally known as the methodology of scientific research programs. Thus, Lakatos attempted to modify what some critics refer to as Popper's "dogmatic falsificationism" by introducing the notion of the "research program." According to Lakatos, an ongoing scientific research program could be divided into three major components: the program's hard core, its negative heuristic, and its positive heuristic. The hard core of a research program consists of a set of theoretical assumptions that determine its specific configuration. The negative heuristic is a methodological proviso stating that the elements in the research program's hard core are not to be abandoned just because of anomalous events. The function of the positive heuristic is to suggest what mechanisms must be appealed to to protect or even salvage the research program's hard core. Hence, in the case of some anomalous event, which threatens the hard core C of some theory T, the positive heuristic would determine what possible hypotheses could be formulated to modify the theory's protective belt, which consists of a set of directly testable auxiliary hypotheses. Thus, a developing successful research program might over time modify its protective belt P to P' (given certain anomalous events) according to the dictates of the program's positive heuristic in order to salvage its hard core C from potential danger. Yet Lakatos's program does not deny the possibility of the ultimate rejection of a scientific theory.

An important aspect of Lakatos's research, therefore, is to point out how theories exist and change through time. In contrast to Popper, he argues that theories are not falsified and rejected as a result of crucial or decisive experiments, and that theorists do not set out deliberately to falsify their theories; they aim to do just the opposite: to confirm and justify the claims of their theories. And against Kuhn's evidently sociological and relativistic approach, Lakatos reintroduces the notion of theory competition and corroboration according to the dictates of scientific research.

It would appear that the methodologies of Popper and Lakatos have struck a more responsive chord among some economists than that of Kuhn. Mark Blaug, for example, argues that "the battle for falsificationism has been won in modern economics.... The problem now is to persuade economists to take falsificationism seriously."[15] But one can imagine why falsificationism is not taken seriously by most neoclassical theorists. Quite obviously, if a falsificationist methodology were applied to neoclassical theory, the results would be quite evident: neoclassical theory would immediately be shown to be a false theory given its predictive and explanatory inconsistencies.[16] It appears, too, that Lakatos's more flexible and less stringent (than Popper) approach to the appraisal of scientific theories and his views on "progressive" and "degenerating" research programs have caught the attention of theorists who are more concerned to examine critically contemporary neoclassical economics given its present problematic status.

Thus, S. J. Latsis would argue that the "protective belt" of a research program in economics is determined by the clash between the theory's "hard core" and its "positive heuristic."[17] According to Latsis, the "hard core" of neoclassical theory consists of no more than the axiomatic assumptions of that theory. But it is at this point that Popper's falsifiability criterion becomes decisive: the fundamental assumptions of neoclassical theory do not make reference to the actual behavior of economic agents; they are merely representative of the idealized behavior of rational economic agents.[18]

Consider, too, Leijonhufvud's attempt to explain the development of Keynesian economic theory, and E. Roy Weintraub's defense of neoclassical general equilibrium theory by appeal to Lakatos's methodology of scientific research programs.[19] I believe that this interest in Lakatos's scientific research criteria derives principally from the fact that the field of economics is methodologically highly unsettled given the nature of the object of its research efforts. Methodologists of economics are concerned to formulate some methodological structure according to which economics as a science or research discipline could be evaluated. Yet paradoxically, Kuhn, Popper, and Lakatos viewed the social sciences and economics in particular as having not yet attained the status of genuine sciences. Their research was concerned essentially to examine the development and career of natural science in history. Perhaps the reason for this is that natural scientists rarely discuss their research in terms of scientific methodology—since that question has already been settled. Possible exceptions could be the quasi-epistemological questions posed by theorists conducting research at the frontiers of particle physics where the freedom to formulate theoretical hypotheses is encouraged given the present constraints on experimental procedures. One could also note the ideologically motivated questions raised by "creation" scientists against the research methods and claims of evolutionary biologists.

Although recent discussion on the methodology of scientific research has been dominated by references to Popper, Kuhn, and Lakatos, it is useful to point to the research efforts of Larry Laudan,[20] who sought to modify the Kuhnian thesis by arguing that since "truth" is too strong a criterion to require of scientific theories, the idea of "problem solving" is more acceptable since it safeguards the notion of "rational inquiry" and instantiates the Kuhnian concept of "puzzle solving." What Laudan offers in contradistinction to Kuhn and Feyerabend is a kind of tempered epistemological relativism. Laudan's thesis states that if some theory T solves some set of problems P in context C, then the scientific tenor of T should be judged as optimal for context C. If, later, T is replaced by T' on grounds of incapacity to solve an increased set of problems P' in context C', then this is not to argue in favor of what one might refer to as the greater truth bearing content of T'. Laudan would argue that both theories are of equal epistemological validity relative to their particular contexts.

I discern a pragmatic or instrumentalist kernel in Laudan's methodology, and it is somewhat surprising that economists and other social scientists have not sought to exploit his ideas given the difficulties encountered by the social sciences in terms of the empirical instantiation of their constituent theories.

I conclude this section on the contemporary methodology of science by stating that economists concerned about the status of contemporary neoclassical economics as a scientific research discipline[21] have expressed much interest in the recent research in the methodology of science, particularly the writings of Kuhn, Popper, and Lakatos. Given the overview of trends in the history and methodology of science, I shall next attempt to formulate some general theory of the structure and proof of an adequate scientific theory based principally on ideas culled from the diverse analyses of scientific methodology.

3
The Structure and Proof of Scientific Theories

The question on what constitutes a scientific theory has been a central one for those concerned with the theory and philosophy of science. Recall in this instance the research efforts of Hempel, Popper, Lakatos, and Laudan. Perhaps the best known definitional effort to date has been that of Popper's, whose criterion of demarcation between science and nonscience is founded on the potential falsifiability of a candidate theory. The approach I take on this issue is one that seeks to pay close attention to the actual methodological practices of the practitioners of science themselves. Thus, I hold minimally that a scientific theory is any structured research program that attempts to explain, predict, and control (if need be) phenomena and events in the world by appeal to repeatable empirical test under appropriately controlled experimental conditions. Note that this definition does not exclude those applied sciences which seek to explain rather than predict since the basic scientific theories on which they are founded are not only explanatory but also predictive. Thus, for example, the most compelling theories in modern biology are those whose structures allow not only for explanation but also prediction. Consider in this regard the stronger experimental warrant of molecular genetics as compared with general evolutionary biology. The former relies maximally on laboratory research in chemistry and physics, while the latter appeals more to hypothesis positing founded principally on logical argument and intuition.

I want to claim now that the structure of any mature scientific theory could be examined in terms of three major components: (1) a conceptual language consisting of observation terms (both direct and indirect) and theoretical terms, (2) deductive algorithms that link together the propositions of the conceptual language, and (3) experimental procedures for purposes of confirmation and possible disconfirmation. It should be pointed out that theoretical structures are not to be viewed as static given the dynamic and often creative nature of scientific research. New concepts may be introduced or rejected according to the dictates of a particular working theory. The same also holds for experimental procedures which are constantly being refined by the most creative of science's practitioners. But consider further analysis of a theory's major components.

The conceptual language of a research science is formulated for the purpose of isolating or bracketing those aspects or portions of the empirical world deemed relevant for analysis. It is for this reason that the conceptual language of quantum mechanics differs from that of, say, molecular genetics. For example, the concept of "photon" and its definition is of crucial importance in quantum mechanics, whereas that of "gene" is not. In the same vein, "gene" and its definition is of central importance in molecular genetics but not in quantum mechanics.

As mentioned above, a theory's conceptual language would contain terms that are generally regarded as theoretical and others describable as observational.[1] However, I should not want to make this sharp distinction since it could be argued that terms that refer to objects that are directly observable are scientifically meaningful only in the sense that such terms refer to objects for which there is confirmable evidence. The distinctions between terms such as *chair*, *molecule*, and *electron* is one of degree rather than kind. It is useful at this point to say something about those terms which describe objects that are not themselves empirically confirmable. Such terms may be viewed as serving a heuristic function on the basis that their usefulness is justifiable only on the grounds that the theories in which they occur prove their mettle instrumentally—in the sense of the theories having evident predictive success. But for purposes of theoretical completeness, heuristic terms must be eventually identified since otherwise the theories in which they occur would be explanatorily incomplete.

I have stated above that the conceptual terms of a scientific theory must contain empirically identifiable content. This requirement is greatly facilitated by describing such terms according to precise measurement techniques. For example, in quantum mechanics there are precise measurement techniques for determining the size and structure of the constituents of the atom. The same concern for precise description, if not quantitative measurement, is expressed by researchers in areas such as molecular genetics and organic chemistry.

The deductive algorithms of a scientific theory serve the function of forging the theory's concepts, propositions and so on into a coherent theoretical whole. For example, the ideal gas law $PV = nRT$ is meaningful within the context of the kinetic theory of gases because its constituent concepts are linked together in the above stated quantitative equational form. The importance of deductive algorithms in scientific research is further evidenced by their role in the formulation of general laws, whose function is to establish causal relationships between phenomena that may not be evidently related. Also, deductive inference in the form of mathematical and quantitative manipulations demonstrate the relationships that constitute the core of a scientific theory.

The requirement that scientific concepts be subject not only to empirical definition but also to precise measurement can be shown to be of major importance in determining whether the concepts of the behavioral and social sciences can achieve scientific status. In the case of economic theory, for example, important concepts such as utility, preference, and rationality do not seem to be amenable to precise empirical definition since they refer to mental states of individuals, which are not subject to empirical observation, hence would be incapable of precise quantitative definition. Hempel, for example, is certainly in error when he argues that theoretical parameters in economic theory such as utility do have the capacity for empirical interpretation, thereby justifying their status as genuine scientific concepts.[2] He seems unaware of the problems associated with the quantification of utility and the theoretical weakness of the ordinal utility theory. It would seem that the strict criticism that Hempel directs against pseudoscientific concepts such as "entelechy"[3] in the theory of neovitalism on the grounds that such concepts are not subject to empirical test are equally applicable to key concepts in economics such as utility and rationality. The point is that scientific concepts that refer ostensibly to empirical objects must demonstrate evident correspondence to relevant portions of the empirical world.

The discussion up to this point would seem to lead to the claim that the ultimate worth of a scientific theory is determined not by the sophistication of its structure but by its capacity to be subjected to experimental tests. Thus, a genuine scientific theory must first specify its mode of application, which would take the form of what may be referred to as an "experimental design." This experimental design would include a "research space" and a set of instruments. The purpose of the experimental design is to isolate that portion of the empirical world relevant to the experiment so that a consensus among researchers concerning the status of the theory could be established.

Accordingly, the status of a scientific theory is determined by whether the results of relevant experiments conform to the claims of the theory. In other words, the theory's predictive claims must be shown to correspond or not with the actual results of experiment. It should be noted that while a test for a correspondence between a theory's predictive claims and its experimental results constitute a necessary condition for the evaluation of its status, it is not sufficient. The theory should be deliberately falsified by the manipulation of its particular variables. It is the combination of the confirmatory and falsifiability potential of a theory that determine its predictive and explanatory capacity.

In fact, it is the researcher's ability to manipulate a theory's experimental design that determines its explanatory capacity—the most important characteristic of a scientific theory. For it is the theory's experimental design that allows the researcher to transform a mere hypothesis into a confirmed or falsified theory. The purpose of scientific explanation, it should be noted, is to authenticate by means of empirical evidence those relevant events and phenomena which yield some particular phenomenon. In those experimental cases where causal phenomena are not directly identifiable, it is the manipulation of a theory's variables that allow the researcher to instantiate a conceptual posit as a causative agent. We establish, therefore, that a genuine scientific theory must be capable of test according to its experimental design so that its predictive and explanatory tenor be established.

It is for this reason that scientific research disciplines such as astronomy and archaeology are best regarded as applied sciences rather than pure sciences, since their researchers are capable of formulating predictive and explanatory statements only in terms of the wholly experimental sciences, namely those sciences in which hypotheses are subject to strict laboratory test. Consider the procedure of determining the age of prehistoric fossils by means of dating by the purely physical test of carbon 14. Note, too, that archaeological research is incapable of confirming predictive statements. It is a purely explanatory discipline but reliant on the predictive and explanatory capacities of more basic laboratory-bound scientific research. Similar considerations apply to astronomy.

Of course, one must indeed continue to pay attention to those theorists who continue to show skepticism about the claims of scientific theories. For example, Nancy Cartwright in her provocative text *How the Laws of Physics Lie*[4] expresses an obvious skepticism about the ontological content of the theories of the natural sciences.

In the case of physics, one can argue that there is much theoretical debate concerning the existence of theoretical entities posited by researchers at its frontiers. But skepticism concerning the status of, say, quarks does not entail an equal skepticism about the structure of the benzene molecule (C_6H_6) or whether photosynthesis is an actual process. The essence of scientific investigation is epistemological skepticism at the frontiers of research, but epistemological warrant of the empirical content of its confirmed theories. Skepticism would always require that a scientific theory stand by its word: does the theory consistently and repeatedly "deliver" what it claims it would deliver?

Scientific Theory and Human Behavior

In the above discussion I have argued that a genuine scientific theory must conform to three criteria: (1) it must possess an appropriate experimental design and a research space; (2) it must possess the capacity to demonstrate whether the theory's explanatory and predictive statements can be empirically instantiated; and (3) a theory's explanatory and predictive capabilities are reinforced by whether its theoretical terms possess empirical content. In fact, the crucial point about a scientific theory is whether it corresponds to sensed events or phenomena in the empirical world.

Given that human behavior is empirically observable, it is thus an obvious candidate for scientific analysis. But unlike researchers of inanimate empirical phenomena, analysts of human behavior must formulate hypotheses whose testability is constrained by the limitations imposed on experimental design by the nature of the object under inquiry. Given that the explanations of human behavior are ultimately reducible to nonempirical mental states such as motives, there is, therefore, the obvious question of the empirical confirmability of such phenomena. Furthermore, the fact that human behavior is not subject to the strict controls of experimental design means that the predictive capacity of theories concerned to analyze it is thereby diminished.

It is for this reason that historically there has been an epistemological divide between those theorists who argue that the nature of human behavior warrants its own special mode of analysis, and those who argue that all empirical phenomena should be interpreted according to a single methodological model. The reference here is to the phenomenologists on one hand and the neopositivists on the other hand. The phenomenologists have argued for a mode of analysis that grasps the intuitive and particular aspects of human behavior, whereas the neopositivists have sought to interpret human behavior as if it were qualitatively similar to the phenomena studied by the researcher in the physical and biological sciences.

It is instructive to note in this connection the foundational arguments put forward by theorists like Bacon, Hobbes, Locke, Hume, Mill, Comte, Marx, and Weber. The unity-of-science approach to the social sciences, no doubt, developed out of the empiricist arguments proposed by Hobbes, Locke, Hume, Comte, Mill, et al. Recall, for example, Hobbes's theory of human behavior founded on the mechanics of Galileo. And Hume's expressed goal in his *Treatise of Human Nature* was "an ATTEMPT to introduce the experimental Method of Reasoning into MORAL SUBJECTS." Hume's unity-of-science approach was, no doubt, modeled on Newton's experimental mechanics. Note, too, that despite the methodological differences between Mill[5] and Comte,[6] their goal was essentially the same in the sense that they were both concerned to establish a science of human behavior. Mill recognized the importance of mental laws because of the causal role they played in explaining human behavior and was indeed willing to argue that there was a basis for a distinct scientific discipline that attempted to analyze purely mental phenomena. Mill would even argue that this new science of ethology (founded on the laws of psychology) was in a more advanced state than the discipline of physiology.[7] But the problem here is that the mental laws that constitute the foundations of Mill's science of human behavior are based on private impressions not subject to public empirical confirmation. Matters are compounded by the fact that Mill argued that the reduction of mental phenomena to physiological states had not been demonstrated.[8]

Mill's methodology was much at variance with that of Auguste Comte, who argued that mental phenomena cannot be viewed as potentially explanatory of human behavior. Comte claimed that mental phenomena, since inaccessible to empirical observation, cannot be viewed as acceptable causes of human choice. Thus, he tended to view the notion of cause as superfluous for a scientifically adequate understanding of human behavior.[9] No doubt, this methodology of scientific investigation was meant to formulate a kind of social physics for the study of human behavior.

But it is evident that a physics of human behavior, though empirically accurate, would be without significance. The reason is that human behavior is meaningful only within a social context. However, meaningful analyses of human behavior are necessarily formulated at the second order level of the subjective interpretation of the observer. Herein lies the dilemma of the social scientist: he or she must attempt to offer meaningful analyses and explanations of human behavior but such analyses must claim objectivity in order to be scientifically acceptable. Matters are compounded by the fact that the basis for the theorist's explanations are mental phenomena that, as seen, are not subject to direct or indirect empirical investigation and public confirmability.

The writings of Max Weber are a good example of the dilemma confronted by social scientists as they seek to interpret human behavior. According to Weber, since "subjective understanding is the specific characteristic of sociological knowledge," explanation founded on this assumption is achieved "at a price—the more hypothetical and fragmentary character of its results."[10]

It is in this context that Weber's important concept of *"verstehen"* or "empathic understanding," comes into play. *"Verstehen"* allows the theorist to offer an explanation for observed behavior in terms of his or her own particular conceptual viewpoint. But note that this method of explanation is highly problematic: the theorist must intuit imaginatively the particular causal mental states of the agents under investigation based on his or her own subjective mental states. The problem derives partially from the fact that the mental states of the observer, though existent, are nonempirical and nonquantifiable.

For future reference, it is useful to note the way in which Weber instantiated the concept of *"verstehen"* for purposes of model construction. To this end, he posited the ideal type notion of "rational behavior" against which behavior that deviated from that model could be compared and analyzed.[11] Although Weber recognized that there was a necessary subjectivity in any analysis of cultural phenomena, he was, nevertheless, of the opinion that a scientifically valid interpretation of cultural phenomena was possible. But it would seem that the presupposition of some explanatory norm of "rational conduct" is scientifically unacceptable. If it were scientifically acceptable to establish a science of human behavior based on the ideal norm of rational conduct, it should then be possible to establish a science of, say, ethical behavior. One would then be able to explain human behavior in terms of how much or how little it deviated from ideally good behavior. The point is that Weber's notion of "empathic understanding" founded on the interpretive concept of "rational behavior" cannot serve the function he claimed it could serve. I will elaborate later on the role of the ideal type "rational behavior" as it is employed in contemporary neoclassical economic theory.

The school of thought developed by Weber and pursued by other thinkers such as Dilthey and Rickert has developed into the contemporary research traditions of phenomenology and hermeneutics. Its adherents claim that naturalistic studies of human behavior do not yield truly significant explanation. But the positivist tradition is still strongly endorsed. Popper, for example, argues that the methods of investigation in the natural and social sciences are basically similar despite evident degrees of difference. The single methodology, according to Popper,[12] consists in appealing to deductive causal explanations, which are to be tested in terms of their predictive capacity. One recalls Popper's methodology of scientific research, which stresses the subjecting of theories to increasingly stringent tests that aim to gauge their falsifiability potential.

A similar unity of science approach is argued for by theorist of science Ernest Nagel. In response to Weber's thesis that subjective interests play a special role in social scientific research while they are not so regarded in the natural sciences, Nagel argues, to the contrary, that a single methodology of investigation is appropriate for both areas of investigation. He writes:

> In short, there is no difference between any of the sciences with respect to the fact that the interests of the scientist determine what he selects for investigation. But this fact, by itself, represents no obstacle to the successful pursuit of objectively controlled inquiry in any branch of study.[13]

However, Nagel's analysis fails to point out an essential difference between the role of "interests" guided research in the natural sciences and the same in the social sciences. Research in the natural sciences requires that the researcher select portions of the empirical world for investigation. But those selected portions of the world must be investigated as objectively and as disinterestedly as possible if the resulting explanations are to be eventually accepted by the scientific community. In the case of the social sciences, the researcher not only selects portions of the empirical world for investigation but also must interpret the human behavior and choices he or she observes in terms of notions that are meaningful only in the subjective sense. This requirement of ignoring the purely physiological or empirical aspects of behavior to focus on its subjective meaning clearly distinguishes research in the natural sciences from that in the social or human sciences. And it is this requirement that forces the social scientist to invoke value judgments in the attempt to interpret and explain human behavior. The free fall of a human body from a given height cannot be meaningfully interpreted in terms of the laws of mechanics only; value-judgmental notions such as "murder," "suicide," and "mental health," must be invoked. It is just these distinctions that Nagel's thesis fails to explore.

Nagel attempts to reinforce his thesis by arguing that the subjective or mental states to which the researcher must ultimately appeal for purposes of explaining behavior have their theoretical counterparts in the natural sciences. Nagel argues, for example, that the ascription of the mental states of fear and hatred as explanatory of a man fleeing from a crowd is similar to the ascription of the increasing velocities of molecules in a piece of wire as an explanation of its increasing temperature. According to Nagel, nonobservable "internal states" reliably explain the observable empirical phenomena.[14]

But, surely, the two cases are not at all qualitatively similar. Knowledge that the rise in temperature of a piece of wire is due to the increase in velocities of its constituent molecules is had by obtaining evidence that the diameter of simple gas molecules is of the order 10^{-9} to 10^{-7} cm, and that electrons and other microparticles (smaller in size than molecules) are observable under appropriate test conditions. On the other hand, the ascription of fear and hatred to the man and the crowd is not subject to similar kinds of proof. The internal states in this case, which correspond to the rapidly moving molecules in the piece of wire, are not the assumed mental states of fear and hatred but the neuronic constituents of brain states. But while brain states are empirically and publicly confirmable, this has not been the case for the psychological states of individual agents. It is just this distinction that prompts theorists of human behavior (following Max Weber) to posit the notion of the "ideal type" of pure rational behavior as a heuristic device to permit the inference from observed behavior to unobserved psychological states.

Yet Nagel himself recognizes the limitations of the concept of the ideal type as a reliable explanatory concept in the human sciences. According to Nagel, it is for this reason that their usage is generally minimal in social-science research where appeal to statistical generalization is the preferred mode of analysis. But it should be pointed out that the correlating of empirical data is best described as taxonomy and is not genuine scientific inquiry. It is the need to explain behavioral phenomena that requires the positing of such heuristic concepts as rational behavior and the like.

Carl Hempel seeks to salvage the usage of the notion of ideal type in scientific research by arguing that ideal types are used not only in the social sciences but also in the natural sciences. Hempel argues, for example, that "ideal gases," "perfectly elastic impact," and so on play roles in physical research similar to that of ideal constructs in the social sciences.[15] The role of ideal types, according to Hempel, is to serve as nonempirical limiting cases against which empirically confirmable cases are measured.[16]

But this procedure would not seem to be followed in the construction of ideal types in the social sciences. It is true that in the natural sciences ideal types may be construed simply as special or limiting cases of some testable theory; however, they are never regarded as model or paradigm cases according to which all explanation is undertaken—which is indeed their role in the social sciences. Again, note that while in the natural sciences ideal types play a limiting role in testable theories, thereby proving that they could be dispensed with altogether, this does not seem to be the case for the social sciences, given the central explanatory role they play. In this connection, Hempel's attempt to establish a unitary theory of science cannot be sustained.

Despite criticisms though, the unity of science and the phenomenological approaches still have their committed adherents. In summary, those who argue against the unity of science thesis claim that social science research cannot measure up to the methodological criteria required of genuine scientific research since the former is necessarily value laden. Human interests, the unstructured nature of experimental design in the social sciences, and the fact that the researcher is both observer and observed are regarded as obstacles that cannot be overcome in practice.[17]

It is interesting to note that this is indeed the methodology practiced by social science researchers though it is usually denied. After all, the subjective value orientations and judgments of social-science researchers do structure empirical reality in particular ways. And the end result to this has been the lack of consensus in the directions and content of theory formulation in the social sciences. Students of the natural and social sciences are, no doubt, struck by the fact of the proliferation of schools of thought in the latter and the general paucity of such in the former. The reason is that natural-science research eschews the normative and aims exclusively at positive empirical results. This is what leads eventually to consensus in theory appraisal. But research disciplines that conflate both the normative and the positive as the social sciences do must necessarily yield intellectual results that would tend to flout the scientific ideal.

It is at this point that one encounters the vigorous ongoing debate as to the role of ideology in the formulation of knowledge. There are those theorists who endorse the traditional views of Marx and Mannheim and argue that interest-derived ideological perspectives determine the way in which theories about the empirical world are fashioned. But what complicates the situation is that these ideological perspectives are both subtly and directly inculcated in the theorist's mode of thinking during the long process of socialization.

The fact that there is less of a problem of ideology in the natural sciences than in the social sciences derives from the notion that usually no special interests are served or injured in the formulation of facts about the empirical world. And given the possible technological benefits that result from the application of natural science theories that accurately describe the empirical world, consensus on the validity and worth of such theories is eventually arrived at. But the claim here is not absolute, given the case of the "Lysenko affair" in the history of Soviet genetics and the growing conflict between those researchers (albeit a small one) who endorse creationism and those who support the tenets of orthodox evolutionary biology.

Kuhn[18] argued that the proliferation of research schools of thought in the social sciences was evidence of their preparadigmatic stages, implying that at some future date they would tend to develop along the lines of the natural sciences. But this would seem rather unlikely as long as societies can be viewed as sociological loci where different interest groups each armed with its own ideology compete for power and the control of resources. This is not to deny that a reasonably objective valuation of society is possible, rather it is to say that as long as social groups have certain interests then the value-free appraisal of social phenomena would be quite difficult. Possible exceptions could be made though for the classless intellectual, in Mannheim's sense of individuals who have not been socially conditioned in any particular systematic way. Such individuals would, no doubt, constitute a very small minority. The best solution, it seems to me, is to advocate the normative position that individuals would eventually gain by accepting those theories which best reflect the empirical world over those which do not. It is in this sense that one could talk of a particular theory in the social sciences being better than another. Obviously, a theory whose basic claims could be shown to be false would be less acceptable than one for which such is not the case. Obvious examples are explanatory accounts in history and sociology. But unlike research in the laboratory bound natural sciences, there is always room for the theoretical rejoinder in the social sciences.

The point is that the epistemological credentials of research in the natural and social sciences are sufficiently distinct to warrant theoretical concern. Despite the fact that theorists of science are much concerned with questions of rationality in scientific research, such questions are not seriously entertained by the practitioners of laboratory-bound scientific research. The empirical and practical yield of the latter has been so impressive that questions of methodology hardly arise in this context. There can be no serious theoretical rejoinders to the atomic theory of matter or the theory of photosynthesis in modern biology. On the other hand, the epistemological difficulties encountered in formulating theories in the social sciences are much in evidence. The inaccessibility of mental states to empirical analysis, the necessary subjective value orientation of the researcher himself and the restrictions on controlled experimental work on human beings makes the formulation of genuine sciences of human behavior a difficult enterprise. One solution would be to discuss ways of modifying the orthodox definition of science, that is, the way in which laboratory-bound natural scientists conduct their research. But that is not what this text is concerned with, since the neoclassical theory of economics—the object of this analysis—claims to be founded on research criteria (minimally, prediction and explanation) consonant with those of orthodox scientific investigation.

A possible critic could find fault with the above approach to scientific methodology on the grounds that discussions on the scientific enterprise by contemporary theorists of science demonstrate that there are many unsettled issues concerning the epistemology of science. Some authors would claim that the present era is an unsettling one of post-Kuhnian epistemological relativism.[19] If one paid attention, however, to the actual efforts of scientists at work in the laboratory, it would be apparent that the concerns of the philosophers are not those of the scientist. I hold the view that science displays cumulative progress in the sense that an increasing number of substantiable claims about the world are being made. I also believe that epistemological relativism derives most of its support from the assumption that the false premodern protoscientific theories of the pre-Galilean era were genuine scientific theories. It would seem that the methodology of research of the pre-Galilean investigator was sufficiently different qualitatively from that of his successor to warrant the claim that both methodologies could not be regarded as being of equal epistemological status. Clearly, alchemy, the Ptolemaic paradigm, and Aristotelian physics (as examples or pre-Galilean "science") are qualitatively more akin to pseudo-scientific theories like astrology than to the research efforts of theorists like Galileo and Lavoisier.

Yet the supporters of epistemological relativism date the beginning of the scientific era with the research efforts of Greek science. I prefer instead to place the birth of genuine science with the period of the Renaissance. The criterion I appeal to is that of logical continuity between theories and their successors. Thus, if some theory B is a successor to another theory A, then both A and B could not be scientific if A were not logically implied by B. This point has been much overlooked by the Kuhnian epistemological relativist who erroneously categorizes as science theories from the period of the development of science, that is, protoscientific theories. It is this conception of natural science as a cumulative discipline that puts it at odds with the research efforts of the social sciences. And this is so even for the "hardest" of the social sciences, psychology.[20]

I conclude this chapter by stating that since the period of the inception of modern science, there have been persistent efforts by theorists of knowledge to determine whether a genuine science of human behavior could be established. The purpose of this chapter was just about discussing these efforts. But my discussion has shown that the success of these attempts has been greatly compromised mainly because of the special and peculiar nature of human behavior. Yet the founders of classical and neoclassical theory were still quite confident about the possibilities of establishing a genuine science of economics. It would seem that the main reason for this is that the early theorists such as Bentham and Jevons believed that if economic decisions could be expressed quantitatively, then the basis for a science of economics would be possible.

Postscript on the Philosophy of Science

A dispassionate student of science and the theory of science would no doubt be struck by the fact that while the enterprise of scientific research is vast and continuously growing with yields confidently presented in the form of novel technologies, the same may not be said for the theory of science. Theoreticians of science seem endlessly engaged in debates as to what constitutes science, with each producing some new approach which, in turn, provokes a set of counter theses in the form of major and minor modifications of some original thesis.

One is witness to theories of realism, antirealism, normative naturalism, epistemic naturalism, conventionalism, epistemological relativism, and so on.[21] One reason for this apparent faddism in the area of philosophy of science (which really has no excuse for this given its relatedness to a firmly grounded and ontologically secure enterprise) is that philosophers of science seem first to develop theories about the nature of science without much regard for the actual activities of scientists.

It is surely the case that modern empirical science informs us about the world of experience and that the success of a particular theory depends on whether it could deliver predictively in clear unambiguous terms on what it promises. The scientist is attracted to a given theory or paradigm according to how that theory's language corresponds to the sensate experiences of the observer. The public whose main contact with the scientific enterprise is by way of its technological applications indirectly accepts a particular theory according to how well a particular technology works. In other words, both scientist and lay individual expect a consistent realization (reinforced by logical rules of explanation) of their preferred theories.

For example, an acceptable scientific theory of a particular disease would be one that identified those individuals afflicted by it, and demonstrated how the cure for it works. There is no substitute for a realist approach. What this means is that an adequate theory of science is one which is also necessarily instrumentalist but strongly reliant on rules of logical inference that link empirically certifiable (technology aided, of course) portions of the experienced world. Were it not for these features, the fruits of the scientific enterprise (consider weaponry, medicine, and so on) would not be so eagerly sought after by all. And even those who harbor doubts about the epistemic status of scientific claims take for granted the technological applications of those claims.

The point I make is that philosophy of science, on account of its subject matter, should seek an epistemology as firm and secure as that of the scientific enterprise itself.[22] The basis for this claim is that the world we live in is a world of experienced phenomena, though it need not have been so. "Confirmation of theories," "truth" or other close synonyms, in the sense of scientific research, are none other than linguistic terms we use to qualify situations in which our anticipations are consistently realized, with no excuses (*ad hoc* hypotheses) accepted. It is this key requirement concerning our anticipations about future states of affairs that is at the heart of the problematic concerning the social sciences and economics in particular.

4
The Classical and Neoclassical Methodology

The success of Newtonian mechanics since its formulation in the seventeenth century was so far reaching that its methodology of research was increasingly viewed as being appropriate for the scientific understanding of human behavior. Recall the influence of Newton's *Principia* on the empiricist orientations of Locke and on Hume's approach to the study of human behavior. In time, the methodology of research in Newtonian mechanics was adopted as the standard for research not only in the natural sciences but also in the social sciences, particularly economics, or "political economy" as it was then called. And what was most important in this regard for the budding social sciences was Newtonianism's appeal to mathematics and the concept of measurement.

This was the basis for the classical economists' subscription to the idea of establishing economic laws and theories that resembled the investigative criteria of classical mechanics. Witness in this regard the early efforts of Smith, Ricardo, Say, and Malthus. Each of these classical theorists sought to explain and predict the operations of the total economy by appeal to basic principles and tendency statements about human behavior. They appealed to formulated propositions about supply, demand and equilibrium, purportedly descriptive of the dynamics of an economic system in much the same way that Newtonian mechanics sought to explain the dynamics of inanimate systems in motion. But a system is composed of its parts, and more specification and measurement were needed to describe more accurately the behavior of the individual units of the economic system.

The specification of this intent may be found in the writings of the two classical utilitarians, Bentham and Mill. The main problem that confronted both writers was to establish some criterion of measurement whereby the empirically observed preferences of economic agents could be explained and predicted. For the utilitarian economists, economic choice was determined by the principle of utility that, in turn, was founded on the twin notions of pain and pleasure. Bentham, for example, writes:

> By utility is meant that property in any object, whereby it tends to produce benefit, advantage, pleasure, good, or happiness (all this in the present case comes to the same thing), or (what comes again to the same thing) to prevent the happenings of mischief, pain, evil, or unhappiness to the party whose interest is considered: if that party be the community in general, then the happiness of the community: if a particular individual, then the happiness of that individual.[1]

And the relatedness of the principle of utility to economics is put by the same author thus:

> Political economy is at once a science and an art. The value of the science has for its efficient cause and measure its subserviency to the art.
>
> According to the principle of utility, in every branch of the art of legislation, the object or end in view is the production of the maximum of happiness in a given time in the community in question.
>
> In the instance of this branch of the art, the object or end in view is the producing that maximum of happiness in so far as the other more general end is promoted by the production of the maximum of wealth and the maximum of population.[2]

Yet what is perhaps most original in Bentham's attempts at establishing a science of economics is his methodology of the quantification of utility: the idea of a calculus of pain and pleasure. It is from this idea of a quantifiable utility that the notion of cardinal utility was introduced into subsequent economic theory. In a section titled "Axioms of Mental Pathology—A Necessary Ground for All Legislative Arrangements,"[3] Bentham proposes an axiomatic approach to the problem of establishing a methodology for positing quantitative statements in economic science.

The subsequent course of the scientific approach to economics was greatly influenced by Bentham's ideas as is proven by the role of utilitarian concepts in the works of the major nineteenth-century economist and utilitarian John Stuart Mill. John Stuart Mill's ideas were shaped basically by James Mill, his father, whose ideas were profoundly influenced by Bentham himself.[4]

It is evident, therefore, that it was generally believed by the economists in the age of a dominant Newtonianism[5] that a major goal of theorists of economics was to establish firm foundations for their discipline by appealing to a proper methodology of scientific investigation. The idea of a quantifiable utility as proposed by Bentham was an important step in that direction.

The social-science counterpart of Newtonianism evolved into what became known as positivism, and it was in this general framework that the key concept of a measurable utility played its all important role. The theoretical foundations were then set for the invention of "homo oeconomicus," that omniscient master calculator of the world of the neoclassical economist. His creators, by virtue of the concept of a measurable utility, were able to determine those choices which would yield him maximum utility.

With measurable utility firmly established as a key theoretical concept in economic science, the stage was set for the marginalist revolution and the birth of neoclassical economics. Historians of economic thought usually discuss the marginalist revolution as being carried out by the individual writings of theorists like Jevons, Menger, and Walras. It was in the theories of this marginalist trio that the microeconomic foundations of economic theory received detailed formal treatment, with discussions centered on the positive and normative aspects of economic theory. It is important to note that the distinction between the positive and normative aspects of economic theory urged by theorists like Pareto and Jevons was not fully endorsed by Walras, who saw no qualitative difference between the descriptive and prescriptive aspects of the discipline. But at the same time, note the groundwork established by the engineer economist Pareto in the development of economics as a science of human choice. He writes:

> All the natural sciences now have reached the point where the facts are studied directly. Political economy also has reached it, in large part at least. It is only in the other social sciences that people still persist in reasoning about words; but we must get rid of that method if we want these sciences to progress.[6]

It was this emphasis on quantitative expression that determined Pareto's development of the key notion of economic equilibrium founded on the idea of maximum ophelimity (utility).

One might recall that the major concern of the classical economists was to determine the value of goods and services and to offer an adequate explanation of this. And questions on value ultimately yielded to questions on the distribution of wealth—questions that are still of major concern today. On the other hand, neoclassical concerns, as expressed in the writings of the marginalist economists, emphasized more the scientific and formal foundations of economic theory. This new emphasis is explainable by the previous writings of theorists like Bentham and Mill who subscribed to the thesis that political economy could be studied scientifically.

The new orientation of the neoclassical economists may be summarized as a shift from economics perceived principally as a discipline concerned primarily with aggregates and policy prescriptions to a discipline concerned mainly with atomistic units and objective statements regarding particular moments in the behavior of these units. The particular moments referred to here are, of course, the moments of the maximization of utility.

The appeal here is to the mathematics of engineering and the utilitarianism of Bentham. Consider in this regard Bentham's approach to economic theorizing as expressed in his *Theory of Political Economy*:

> In this work I have attempted to treat Economy as a Calculus of Pleasure and Pain, and have sketched out, almost irrespective of previous opinions, the form which the science as it seems to me must ultimately take. I have long thought that as it deals throughout with qualities, it must be a mathematical science in matter if not in language. I have endeavoured [sic] to arrive at accurate quantitative notions concerning Utility, Value, Labour [sic] Capital, etc., ... The theory of economy, thus treated, presents a close analogy to the science of Statical Mechanics, and the Laws of Exchange are found to resemble the Laws of Equilibrium of a lever as determined by the principle of virtual velocities.[7]

In the context of the discussion, Jevon's main contribution to economic theory lies in his theory of utility and theory of exchange. By appealing to the mathematical tool of the differential calculus, Jevons was able to express the notion of marginal utility and exchange in strictly quantitative fashion. He also established the "law of the variation of utility," expressible both in graphical and mathematical form. Of obvious familiarity, too, is the notion that at equilibrium the marginal utilities derived from different commodities are equal. Jevons writes:

> Hence when the person remains satisfied with the distribution he has made, it follows that no alteration would yield him more pleasure; which amounts to saying that an increment of commodity would yield exactly as much utility in one use as in another. Let u_1, u_2, be the increments of utility, which might arise respectively from consuming an increment of

commodity in two different ways. When the distribution is completed, we ought to have $\triangle u_1 = \triangle u_2$; or at the limit we have the equation

$$\frac{du_1}{dx} = \frac{du_2}{dy}$$

which is true when x, y are respectively equal to x_1, y_1. We must, in other words, have the *final degrees of utility* in these two uses equal.[8]

Other contributions of Jevons include the theory of exchange derived from the law of diminishing marginal utility. Jevons expressed this equation of exchange thus:

$$\frac{\phi(a-x)}{\psi_1 y} = \frac{y}{x} = \frac{\phi(x)}{\psi_2(b-y)}$$

where a and b represent two commodities held by two parties and x and y represent the quantities exchanged and ϕ and ψ represent the final degree of utility to the two parties. The modern version of this equation is:

$$\frac{MU_x}{MU_y} = \frac{P_x}{P_y} \quad \text{or} \quad \frac{MU_x}{P_x} = \frac{MU_y}{P_y}$$

The writings of Walras are also of significance in the persistent attempts of the early neoclassical economists to develop a legitimate science of economics. Walras himself endorsed the theories expounded by Jevons and took for granted the latter's statements on the solution to the problem of the quantification of utility. Walras's importance in the theory of neoclassical economics derives from his attempt to establish a general and formal theory of equilibrium for the economy as a whole. The significance of his writings is such that some comment thereon is warranted.

Note first of all that one finds in Walras's writings the familiar attempts at establishing the science of economics on firm ground by utilizing a theoretical infrastructure and methodology that were similar to those of the natural sciences. Walras writes:

This much is certain, however, that the physico-mathematical sciences like the mathematical sciences, in the narrow sense, do go beyond experience as seen as they have drawn their type accounts from it. From real-type concepts, these sciences abstract ideal type concepts which they define, and then on the basis of these definitions they construct a *priori* the whole framework of their proofs and theorems. After that they go back to experience not to confirm but to apply their conclusions. . . . Following the same procedure, the pure theory of economics ought to take over from experience certain type concepts, like those of exchange, supply, demand, market, capital, income, productive services, and products. From these real-type concepts, the pure science of economics should then abstract and define ideal type concepts in terms of which it carries on its reasoning. The return to reality should not take place until the science is completed and then only with a view to practical applications. Thus, in an ideal market we have ideal prices which stand in exact relation to an ideal demand and supply. And so on.[9]

This confident position is easily explained by the following statement in Walras's preface to the fourth edition of the *Elements of Pure Economics:*

In any case, the establishment sooner or later of economics as an exact science is no longer in our hands and need not concern us. It is already perfectly clear that economics, like astronomy and mechanics, is both an empirical and rational science.[10]

In an attempt to establish a complete and general theory of economic activity, Walras formulated a theory of general equilibrium from which was deduced the following important theorem:

The theorem of general equilibrium in the market may be stated in the following terms: When the market is in a state of general equilibrium the $m(m-1)$ prices which govern the exchange between all possible pairs drawn from m commodities are implicitly determined by the $m-1$ of these commodities and the mth. Thus, the situation of a market in a state of general equilibrium can be completely defined by relating the values of all the commodities to the value of any particular one of them.[11]

One curious point though, is that Walras did not seem too much concerned with the problem of the actual measurement of utility. Perhaps he assumed that such problems were already settled by Jevons and others, for one may note that his notion of the maximization of utility is simply stated as "when the ratio of the intensities of the last wants satisfied, or the ratio of their *raretés* is equal to the price." And, although "*rareté*" is defined as "intensity of the last wanted satisfied," Walras is not concerned to give a further definition of intensity. The term "intensity" would seem to imply subjective states that may not be amenable to quantitative expression.

In the discussion so far, it has been pointed out that the development of economics as an area of scientific inquiry went hand in hand with the assumption that "utility," the major determinant of economic choice, was capable of being measured. Recall that this particular notion of utility was introduced into economic literature by Bentham. And the idea of a measurable utility was accepted without much question by such important neoclassical authors as Jevons, Menger, and Walras. This notion of utility was later adopted and modified by Marshall. What is significant though, is that the early neoclassicists all assumed the idea of an empirically measurable or additive utility. The utility of a commodity was a function of the quantity of that commodity, which was independent of other commodities consumed.

The ideas put forward by the early neoclassical theorists on questions concerning the status of economics as an empirical science and the way in which the important concept of utility was subjected to measurement were developed further by Marshall, thereby giving neoclassical economics the stamp of maturity. The important question of the logical distinction between scientific economics and evaluative or welfare economics was assumed to be settled by the latter, thus endorsing the theoretical assumptions of his predecessors. Marshall writes:

> Lastly it is sometimes erroneously supposed that normal action in economics is that which is right morally. But that is to be understood when the context implies that the action is being judged from the ethical point of view. When we are considering the facts of the world, as they are, and not as they ought to be, we shall have to regard as "normal" to the circumstances in view, much action which we should use our utmost efforts to stop.[12]

The science of economics also "shuns many political issues which the practical man cannot ignore: and it is therefore a science, pure and applied, rather than a science and an art."[13]

It is evident then, that a mature neoclassical economics was firmly wedded to the idea of a measurable utility as a necessary condition for its scientific status. It will be seen further on in the discussion how this notion of cardinal utility gave way to the notion of choice-ordering, thereby ushering in a novel phase in the history of economics.

One of the important implications of the idea of cardinal utility was that utility was additive. And this obviously led to the measurement of individual choice making in terms of the intensity of choice. This characteristic of the cardinal measurement of utility is generally referred to by economists as "utility measurable up to a linear transformation." Thus, granted the assumption that some consumer confronted with three commodities A, B, and C ranks them cardinally as $A > B > C$ in terms of the utils of satisfaction derived, *two* important statements could be made about this ranking: not only that $U_A > U_B$ and $U_B > U_C$, but also that there is a determinable quantitative comparison of the preferences in question. It is possible, therefore, to say that $U_A > U_B$ is thrice $U_B > U_C$. The virtue of the cardinal interpretation of utility derives from the fact that it permits utilization of the differential calculus. The idea of comparing differences leads to talk not only of first differences but also of second differences. First differences signify the order of the ranking of the preferences, while second differences signify the intensity of preference. And calculus operations offer a mathematical interpretation of the law of diminishing marginal utility by the negative sign of the second derivative; the first derivative, of course, offers a measurement of marginal utility.

But the actual measurement of utility in terms of intensity of preferences is indeed problematic, since the measurement of the intensity of preference of X over Y requires also a measurement of such between Y and Z. This works, though, only so long as the utility of one good is strictly independent of that of any other good. If this independence criterion does not hold, new utility functions not linearly related to an old function will result whenever there is some change in its components. Thus, for example, it will be impossible to have recourse to cardinal utility measurement by means of pairwise choices if $U_X = f(X)$ and $U_B = f(B)$ were replaced by the generalized function $U_X = f(X, Y, Z, \ldots)$.

One further point, though, is that the principle of additive utility does not permit the existence of inferior goods—although it is an empirical fact that over certain income ranges such goods do exist. But Marshallian economic theory opted for the additive and independent utility thesis on the basis that it is difficult to devise an easily manipulable method for measuring cardinal utility using a generalized utility function. Furthermore, recourse to a generalized utility function makes it quite difficult to establish downward sloping demand curves and upward sloping income curves necessary for analysis.

The strength of the cardinal approach to the measurement of utility, firmly established by Marshall after the groundwork was laid down by Jevons and others of the neoclassical school, is based on the fact that by assuming the constancy of the marginal utility of money the measurement of utility unique up to a linear transformation was made possible.

Perhaps one of the most notable efforts to establish an adequate theory of cardinality, in face of questions of methodology concerning utility measurement and the introduction of the alternative ordinal theory of utility, is that of J. von Neumann and O. Morgenstern who, in the *Theory of Games and Economic Behavior*,[14] attempt to derive a cardinal utility index applicable to consumer behavior under conditions of risk. The authors declare that "treating utilities as numerically measurable quantities—is not as radical as is often assumed in the literature" (p. 16). And "that under the conditions on which the indifferences curve analysis is based very little extra effort is needed to reach numerical utility" (p. 17).

Consider the following key statement:

> We expect the individual under consideration to possess a clear intuition whether he prefers the event A to the 50–50 combination of B or C or conversely. It is clear that if he prefers A to B and also to C, then he will prefer it to the above combination as well; similarly, if he prefers B as well as C to A, then he will prefer the combination, too. But if he should prefer A to, say B, but at the same time C to A, then any assertion about his preference of A against the combination contains fundamentally new information. Specifically: if he now prefers A to the 50–50 combination of B and C this provides a plausible base for the numerical estimate that his preference of A over B is in excess of his preference of C over A. If this standpoint is accepted, then there is a criterion with which to compare the preference of C over A with the preference of A over B. It is well known that thereby utilities—or rather differences of utilities—become numerically measurable.[15]

Recall that for von Neumann and Morgenstern the idea of preference intensity is not derivable from the notion of risk, but from the notion of the consumer expressing quantifiable preferences between some choice and other probabilistically determined choices. The von Neumann–Morgenstern theory suggests that some numerical estimate represents the "ratio of the preference of A over B to that of C over B."[16] Thus, for some situation in which the intensity of preferences between A and B equals the intensity between C and A, we have $U(C) - U(A) = U(A) - U(B)$. And $2U(A) = U(B) + U(C)$. Thus, $U(A) = \frac{1}{2}U(B) + \frac{1}{2}U(C)$.

The above indicates that comparisons of preference intensities are equivalent to the notion that the individual's choices in risk situations are geared toward the maximization of expected utility. But the von Neumann–Morgenstern attempt to reintroduce the notion of measurable utility in economic theory has been questioned by some theorists[17] who argue that although this theory introduces the important idea of consumer choice under conditions of risk, it has not in any way solved the problem of the measurability of preference intensity between choices. This cannot be otherwise, since the theory does not relax the assumption that interpersonal comparisons of utility are not possible.[18] And this is indeed a serious matter since the cardinal measurement of utility is possible only if there exists some measuring rod that could accurately measure individual utilities, just as one would measure the heights and weights of individuals. The problem, of course, is that the "utils," supposedly measurable by the cardinal utility mechanism, are not publicly observable phenomena such as height or weight but are rather *privately* generated and experienced. This is the basis for the assumption that interpersonal comparisons of utility are not theoretically acceptable. Yet on the other hand, the cardinal utility theory is not really viable without the possibility of interpersonal comparisons of utility.

The result of the von Neumann—Morgenstern restriction on interpersonal comparisons of utility is that both authors seek to formulate their theory on the same basic axiomatic assumptions of the rival ordinal utility theory. For example, the crucial axiom of transitivity is retained and plays a key role in their theory.

Let me summarize this important theory as follows. The von Neumann—Morgenstern theory introduces into modern economic theory the notion of choice making under conditions of uncertainty. It attempts to quantify choices made within the context of ordinal utility theory, but the theory's inability to establish a model whereby interpersonal comparisons of utility are possible demonstrates the same weakness that led to the replacement of the classical cardinal theory by the ordinal theory. The attempt to measure utility in the same way that theorists in the natural sciences measure heat, mass, and so on seems questionable on the ground that while concepts in physical science are measured according to interpersonal indices of measurement, there is no equivalent index of measurement applicable to economic choice.

Yet the influence of the von Neumann–Morgenstern theory on modern neoclassical theory is evident in contemporary expected utility theory. Expected utility theory, though concerned only with ordinal rankings between risk preference alternatives, owes much to the von Neumann–Morgenstern appeal to probability theory as a means of formulating propositions of decision making under conditions of risk.[19]

Yet in terms of the history of economic methodology, the problem of the measurability of utility as demanded by the cardinalist approach continued to be a source of bother for the post-Marshallian economics. Admittedly, the cardinalist approach was based on subjective notions of pleasure or satisfaction—which indeed were not subject to direct empirical observation. The fact that interpersonal comparisons of utility were ruled out for consumer behavior analysis implied that no empirically certifiable common *index* of measurement had yet been devised to allow the measurement of utility. Consider, for example, the early observations of Lionel Robbins that:

> It is one thing to assume that scales can be drawn up showing the *order* in which an individual will prefer a series of alternatives, and to compare the arrangement of one such individual scale with another. It is quite a different thing to assume that behind such arrangements lie magnitudes which themselves can be compared.[20]

In the same context, Paul Samuelson, a contemporary theorist, writes:

> It is more than half a century since the first formulations of utility analysis by Jevons, Menger, and Walras. In that time, there has been much controversy for and against this concept.
>
> Although much of the discussion has not yet gone beyond a quasi-philosophical defense or rejection of the utility concept, it is nevertheless possible to discern clear lines of development in the literature. First there has been a steady tendency toward the removal of moral, utilitarian, welfare connotations from the concept. Secondly, there has been a progressive movement toward the rejection of hedonistic, introspective, psychological elements. These tendencies are evidenced by the names suggested to replace utility and satisfaction—ophélimité, desirability, wantability, etc.[21]

The solution to the problems of cardinal measurement was then the adoption of ordinal utility theory and the development of indifference curve analysis. Indifference curve analysis was invented by F. Y. Edgeworth, then developed by W. Pareto, I. Fisher, and J. R. Hicks and R. G. D. Allen.

The main point at which the ordinal approach to the measurement of utility differs from the cardinal approach is that while the cardinal approach yields functions "unique up to a linear transformation," the ordinal approach yields functions only "unique up to a monotonic transformation." In other words, ordinal analysis reveals only that $U_A > U_B$ and that $U_B > U_C$. Questions as to intensity of preference (answerable in the cardinal utility approach) are not determinable within the ordinal utility context. Furthermore, the idea of diminishing or increasing marginal utility has no meaning. Ranking by order of preference yields questions about marginal utilities, but only as to whether these utilities are positive or negative.

But the ordinal approach differs from the cardinal aproach in an important methodological way. The traditional Benthamite idea of utility measurement founded on the introspective notions of pain and pleasure finally gives way to the strictly behaviorist ranking or preference approach. In other words, utility is no longer determined by the amount or intensity of pain or pleasure derived from an object; it is now determined by empirically determined order or rank of choice. Thus, consumer theory has been shorn of excess baggage in the form of psychological notions not subject to empirical interpretation. As Hicks himself succinctly put it:

> We have now to undertake a purge, rejecting all concepts which are tainted by quantitative utility, and replacing them, so far as they need to be replaced, by concepts which have no such implication. The first victim must evidently be marginal utility itself. If total utility is arbitrary, so is marginal utility. But we can still give a precise meaning to the ratio of two marginal utilities, when the quantities possessed of both commodities are given.[22]

But it is Hicks's "second victim" that represents what one might call a genuine "paradigm shift." He continues:

> The second victim (a more serious one this time) must be the principle of Diminishing Marginal Utility. If marginal utility has no exact sense, diminishing marginal utility can have no exact sense either. But by what shall we replace it? By the rule that the indifference curves must be convex to the axis. This may be called in our present terminology, the principle of Diminishing Marginal Rate of Substitution.[23]

Hicks then goes on to state that the theory "need(s) the principle of diminishing marginal rate of substitution for the same reason as Marshall's theory needed the principle of diminishing marginal utility."[24] But it was Samuelson who completed the formulation of the new ordinal theory of utility by taking issue with the Hicks-Allen notion of the marginal rate of substitution. Samuelson writes:

Hence, despite the fact that the notion of utility has been repudiated or ignored by modern theory it is clear that much of even the most modern analysis shows vestigial traces of the utility concept. Thus to any person not acquainted with the history of the subject, the exposition of the theory of consumer's behavior in the formulation of Hicks and Allen would seem indirect.... I propose, therefore, that we start anew in direct attack upon the problem dropping off the last vestiges of the utility analysis.[25]

The last "vestiges of the utility analysis" to which Samuelson refers are "the marginal rate of substitution," regarded as "ambiguous" and "an artificial convention in the explanation of price behavior."[26] It was on the basis of this attempt to secure a firm empirical grounding for economic theory that Samuelson proposed the theory of revealed preference. The original intent of revealed preference theory was to obtain the same results derived from the new ordinal utility theory without recourse to any unnecessary nonempirical terms. Samuelson researcher Stanley Wong has argued, for example, that questions could be raised as to the significance or actual content of the revealed preference theory.[27] I shall explore this claim more fully in the following chapter.

In sum it may be said that from an initial starting point of positing a quantitative basis for the measurement of utility, economic theory has progressed to the point where an ordinal concept of utility has led, in some quarters, to the replacement of the differential calculus by the axiomatic choice-theory approach. One might recall that differential calculus usage was introduced to offer a quantitative interpretation of diminishing marginal utility. Further developments in the axiomatic choice-theory approach focus on problems of choice under conditions of uncertainty. It should always be borne in mind, though, that the changes that have taken place in economic methodology have been due to a persistent effort on the part of theorists to construct a firm scientific basis for economics. This scientific basis would exclude all normative assumptions and would permit only the strict explanation and prediction of economic phenomena.

But for purposes of discussion, it is instructive to point out that the belief that the theories of economics were firmly grounded on empirical data was not accepted by all theorists. For example, T. W. Hutchison argued that most of the basic propositions of economic theory, including those qualified by the *ceteris paribus* assumption, were, in principle tautologous since they did not lend themselves to empirical test.[28] Notable responses to Hutchison's skepticism include those of Machlup, who argued against the notion that the fundamental postulates of economic theory could be unambiguously qualified as either analytic (tautologous) or synthetic (capable of empirical confirmation). For Machlup, the fundamental propositions of economic theory should be evaluated heuristically as propositions that afford a useful framework for the analysis of economic phenomena.[29] It is evident from this and subsequent comments on methodology made by both authors that they differ radically on the kind of confirmation required of the basic propositions of economic theory.[30]

But the useful point to recognize is that in spite of the stimulating positivist and post positivist debates, it was the neoclassical paradigm that promoted the scientific credentials of economic theory: the rejection of the notions of cardinal utility in favor of ordinal utility, increasing quantitative expression, and the positing of theories of revealed preference as fundamental for microeconomic decision making.

Furthermore, there was much support from the neopositivist philosophers of science. Although there is the recognition that the ultimate basis for economic decision making derives from subjective mental states, philosopher of science Nagel, in an assessment of the methodology of the social sciences, writes:

> The crucial point is that the logical canons employed by responsible social scientists in assessing the objective evidence for the imputation of psychological states do not appear to differ essentially (though they may often be applied less rigorously) from the canons employed for analogous purposes by responsible students in other areas of inquiry.[31]

The position stated above does not, however, fully resolve the problem. It is a fact that one can imaginatively ascribe motive or feeling states to agents in choice-making situations, but as the critique of the concept of cardinal choice demonstrates, imaginative ascription of feeling states is not subject to empirical verification or quantitative analysis. On the other hand, researchers in the natural sciences are not required to offer explanations for the behavior of phenomena under investigation in terms of the subjective feeling states of such phenomena. It is surely the case that molecules do not make conscious decisions to behave in one way rather than another.

Consider, too, the views of physicist and theorist of science Henry Margenau, who argues that there is no qualitative difference in the way in which theoretical constructs are formulated in both the natural and social sciences. Margenau has formulated the notion of a theory as comprising a P-field containing immediately sensed or protocol experiences, and a C-field consisting of constructs fashioned out of the protocol experience. The transition from the P-field to the C-field is effected in similar fashion for both natural and social science. In the case of economics, Margenau argues that:

> The crude facts [protocol data] of personal economics are translated into objective constructs by very specific operational definitions which make these protocol facts objective, meaningful, quantifiable and subject to logic and mathematics.[32]

Yet I would want to argue that the transition from a P-field to a C-field as evidenced in the natural sciences is not really repeatable as in the case of economics. I accept the claim that observation of "scintillations on a zinc-sulfide screen" entails the assertion that "electrons exist." However, it is difficult to accept the inference from the empirical observation of economic behavior to non-empirical assumptions that guarantee choices about the maximization of utility or profits.

In the above chapters, I have been concerned to examine the question of the methodology of science as it applies not only to the natural sciences but also to the social sciences. In this chapter further progress was made toward the object of analysis, that is, the neoclassical economic theory, by discussing the shaping influence of a maturing Newtonian paradigm on a nascent neoclassical economic theory. The influence of the Newtonian paradigm is evident from the structure of the evolutionary path taken by the theorists of economics from the classical period of Adam Smith and his successors through J. S. Mill up to the period of the neoclassical formulations of Jevons, Menger, and Walras. Note the transition from the purely verbal though systematic analyses of Smith and Ricardo to the calculus (of Newtonian and Leibnizian origin) based formulations of Jevons and Walras. In this context, what was then known as political economy gives way to economic science, a discipline committed principally to the purportedly objective study of economic phenomena.

Thus, the model for neoclassical economics was that of Newtonian mechanics: initially the graphical analysis of early Newtonian mechanics then the calculus approach of a mature mechanics. But the most interesting aspect of this adaptation was the substitution of the idea of an artificial "economic man" for the inanimate Newtonian body in motion. While the latter's behavior was determined by Newton's laws of motion, the former was motivated by the notion of utility and profit maximization. The heuristic role of the construct "economic man" was that it allowed for the prediction and explanation of choice within the neoclassical model.

And finally, since Newtonian mechanics greatly emphasized the measurement of phenomena, the neoclassical theory required the measurement of utility, the basis of economic choice. Herein lies the origin of the cardinal utility theory. But as seen, "cardinal utility" proved to be a problematic concept given that real economic man was quite different from his artificial homologue. The point is that the Newtonian model applied only to inanimate phenomena (recall the application of Occam's razor to the medieval natural philosophy concepts of *vis viva*, sympathy, etc.), while real economic man was not only animate but a creature of mind and body, preference and choice, and capable of fallible calculation. The problem of the measurability of utility is at the base of the paradigm shift from the cardinal to the ordinal theory of utility. I shall explore the implications of this in the following chapter.

5
Ordinal Utility Theory and Contemporary Neoclassical Economics

The critique of the cardinal utility program undertaken by Hicks and Samuelson was so successful that the axiomatic foundations of contemporary neoclassical economics are now assumed to be founded exclusively on the ordinal utility theory.[1] Having given up the possibility of utility measurement, the theorists of ordinal utility have sought to imbue that theory with empirical content by deriving the theory of revealed preference from it.

However, this discussion will show that the ordinal theory is beset with methodological problems as important as those which affected the cardinal utility theory. I shall be concerned first to examine the structure of the revealed preference theory and to comment on criticisms of it. I shall also see fit to formulate the key structural elements of the neoclassical theory and to discuss the various methodological defenses marshalled in its favor. It should be noted that the material to be cited is standard and familiar to textbook readers of microeconomic theory, but in this context it would be examined with the following questions in mind. What is its relevance to the actual behavior of individual agents? Should the mathematically expressed choice paths of the consumer or entrepreneur be viewed rather as prescriptive programs for an artificial *homo oeconomicus?*

What will be evident from this analysis, however, is that contemporary economic theory is founded on two distinct methodological structures. On the one hand, there is the ordinal theory founded exclusively on a more recent choice and decision theory.[2] On the other hand, there is the older differential calculus approach of the theories of consumer behavior and the firm. The function of this approach is to determine in quantitative terms the agent's maxima in terms of consumption, profits, and so on. But in the maximization of utility, the theory of cardinal utility is still very much at work, theoretical disclaimers notwithstanding. It would seem that there is no logical connection between these two approaches. Similar theoretical concerns may be expressed against the formulation of the theory of choice in situations involving risk.

Methodology and Revealed Preference Theory

Revealed preference theory is generally credited to Samuelson and Houthaker with their respective formulations of the weak and strong axioms of revealed preference. The weak axiom of revealed preference states the following: if q^1 is revealed to be preferred to q^2, the latter must not subsequently be revealed preferred to q^1, or that $p^1 \cdot q^2 \leq p^1 \cdot q^1$ implies that $p^2 \cdot q^2 < p^2 \cdot q^1$ for $q^2 \neq q^1$. Houthaker's strong axiom of revealed preference may also be stated as follows: if $q^1, q^2, \ldots, q^n$ are commodity bundles, of which at least two are distinct, and if $p^1, p^2, \ldots, p^n$ are the associated prices then $p^1 \cdot q^2 \leq p^1 \cdot q^1$, $p^2 \cdot q^3 \leq p^2 \cdot q^2, \ldots, p^{n-1} \cdot q^n \leq p^{n-1} \cdot q^{n-1}$ implies $p^n \cdot q^n < p^n \cdot q^1$ for all integers $r > 0$.

First, despite Samuelson's original attempt to formulate a theory of agent choice free from utility considerations, Houthaker has demonstrated the logical equivalence between ordinal utility theory and revealed preference theory. This is to be expected since revealed preference theory and ordinal utility theory share important features, such as the postulate of rationality and its constituent axioms. It would seem that if two theories are logically equivalent, then their ontological claims would be similar. But this could not be possible if the revealed preference theory possessed more empirical content than the ordinal theory. I shall demonstrate later in this text that the ordinal theory, given its reliance on the postulate of rationality, does not possess genuine empirical content. Thus, if the revealed preference theory and ordinal theory are logically equivalent, then the claim by Henderson and Quandt, two prominent textbook theorists, that the theory of revealed preference allows the prediction of agent behavior is subject to question.[3] It should be evident then, that any epistemological critique of the ordinal theory applies equally to the theory of revealed preference.

The view that revealed preference theory adds nothing in terms of empirical content to ordinal theory is maintained by Blaug[4] and Wong.[5] Wong's detailed analysis of Samuelson's revealed preference theory yields the following noteworthy observations:

> First, revealed preference theory, as revised by Houthakker, is not an explanation but a restatement of ordinal utility theory. Second, revealed

preference theory is not verifiable empirically because it uses unrestricted universal statements. Third, it is not verifiable empirically because its key term, "revealed preference," is not defined exclusively in observational terms, and does not therefore denote observable experience, and because the terms "price" and "quantity" are not pure observational terms.[6]

I concur with Wong on his first and third points above, but his second claim (supported by Blaug) is problematic. Revealed preference theory may be unacceptable as genuine scientific theory, but it is not because it uses unrestricted universal statements. On the contrary, successful scientific theories do rely on unrestricted universal statements, usually referred to as general laws, for purposes of explanation and prediction. I believe that Wong's claim in this instance derives from Karl Popper's well-known methodological thesis that universal statements cannot be verified on account of logical problems connected with inductive inference. But Popper also argued that while universal statements are not subject to verification, they could be falsified. Popper's claims on the topic of hypothesis verifiability and falsifiability may have been too strict, since scientists not only reject hypotheses but also accept them.

The question then is what term should be used in describing those hypotheses or general statements which have consistently proven their validity and worth over time? Can one argue that the well-known hypothesis of photosynthesis in biological science is verifiable? In any case, methodologists of science influenced, no doubt, by Popper prefer to speak of the "confirmation" or "corroboration" of hypotheses rather than their "verification." Apparently, the term "confirmation" carries less of a sense of finality and certitude than "verification." I have doubts about this, though, since "confirmation" may be regarded as a lexical synonym for "verification." But if one should split linguistic hairs and choose to speak of the "confirmation" of hypotheses rather than their "verification," then one should be equally willing to reject usage of the phrase "falsification of hypotheses" in favor of "disconfirmation of hypotheses." After all, if it is permissible to argue that no number of confirming cases of the biological science hypothesis of photosynthesis could make that hypothesis verified, then analogously, no number of disconfirming cases could make it falsified. No doubt, this claim could prove to be somewhat disconcerting to those who support the notion that science is intrinsically a progressive enterprise.

Examination of the neoclassical microeconomic theory will begin, therefore, with the analysis of the theory of agent behavior as founded on the concept of ordinal utility. One notes immediately that the fundamental axioms of consumer theory do not derive from the empirical observations of the choices of sets of consumers but on the specific idealized preferences of "rational economic man." As suggested elsewhere, the reason why the neoclassical theorists were obliged to make this theoretical choice is that the analysis of economic behavior does not entail recourse to the strict controls of the scientific laboratory. These restrictions have led to the fact that genuine laws of human choice making have not yet been formulated. Furthermore, the historical attempts to explain overt human behavior patterns in terms of nonempirical psychological laws have not been successful, as noted. I declare once more that cardinal utility theory failed on account of this theoretical impasse.

It is acknowledged, therefore, that contemporary neoclassical economic theory is founded on the premises of the ordinal utility theory. It is the ordinal theory that serves as the basis for the current orthodox theories of consumer and entrepreneur behavior. In the following sections, it is necessary to set down the essential structural components of contemporary neoclassical theory in order to analyze them fully.

The Axiomatic Structure of the Ordinal Theory

Consider now the basic axiomatic structure of contemporary neoclassical theory. Note first that the choice space X and Y of the neoclassical agent as consumer and entrepreneur, respectively, constitutes the nonnegative Euclidean n-orthant and that X and Y are closed and convex. Since I shall focus on the choice paths of the neoclassical agent as consumer in this section, assume that he or she (j) possesses a utility function and is presented with a set of nonnegative prices and a certain amount of income. Thus, one can define the consumption set of j as $\overline{X}_j$ where $\overline{X}_j$ is a proper subset of X. Furthermore, $x^j \varepsilon x_j$ is defined as any particular individual commodity bundle in the subset X_j. One also assumes that $Z = [X, Y]$ stands for any situation of the economy descriptive of the choice patterns of individual agents whether consumers or producers. Assume too, that Z stands for any set of states of the economy.

Neoclassical ordinal theory now requires that a few basic ranking predicates be posited for purposes of describing the behavior of the rational agent. Thus, "P" stands for "strictly preferred," "R"

stands for "is at least as preferred as," and "I" stands for "is indifferent to."

The theory may now be structured according to the following definitions and axioms, which describe the preferences and choice paths of the rational neoclassical agent.

Definition 1. If Z', $Z'' \varepsilon S$ and $Z'P_jZ''$ then not $Z''R_jZ'$

Definition 2. If Z', $Z'' \varepsilon S$ then $Z'I_jZ''$ means $Z'R_jZ''$ and $Z''R_jZ'$.

Completeness Property: If Z', $Z'' \varepsilon S$, then $Z'R_jZ''$ or $Z''R_jZ'$.

Transitivity Property: If Z', Z'', $Z'''\varepsilon S$, and if $Z'R_jZ''$ and $Z''R_jZ'''$ then $Z'R_jZ'''$.

The transitivity property has been a source of concern for some theorists, no doubt, because it has been pointed out that decision makers have been observed to make intransitive choices. The purpose of the neoclassical model, though, is to ensure that the rational agent makes consistent choices that could be guaranteed by requiring, instead of transitivity, acyclycity of choice or convexity of preferences.[7] But given that the utility function still plays an important role in orthodox agent choice, there is some question as to whether the required assumption of strong convexity of preferences does not implicitly subsume transitivity of preferences.

Reflexivity Property: If $Z'\varepsilon S$, then $Z'R_jZ'$.

Asymmetry: If Z', $Z''\varepsilon S$, then $Z'P_jZ''$ implies not $Z''P_jZ'$.

Continuity: In order to prevent discontinuities in the agent's preferences, the assumption is that $[x' | x'\ R\ x]$ and $[x' | x\ R\ x']$ are closed sets for every commodity bundle x. The theory thereby rules out lexicographic orderings.

The following assumptions are also posited by the theory:

(1) There are no bliss points such that the incremental con-

sumption of any item leads to a fall in utility. This may be expressed formally as: there is some commodity bundle $x^j \varepsilon X_j$ and some other bundle $y^j \varepsilon X_j$ with $|x^j - y^j| < a$ such that $y^j P_j x^j$.

(2) Strong convexity of preferences: In order to prevent concave or flat indifference curves, the theory argues that a set of choice elements, among which the agent is indifferent when expressed as a contour map, is strongly convex to the origin in positive n − space. The theory requires that if $x'^j I_j x''^j$, then for all $0 < a < 1$ $[ax'^j + (1 - a)x''^j] P_j x''^j$.

(3) The agent possesses a quasi-concave utility function, which allows him to rank his choices ordinally. These choices are determined by the prices of objectives and the agent's income constraints.

Structural Elements of the Neoclassical Model

In this section, I proceed to offer an operational description of the neoclassical model. However, what follows is just a sketch of the standard formulations of the latter since only the most important aspects of ordinal utility theory will be presented. Consider in this regard, therefore, the following standard assumptions of agents' profiles and choice paths.

(i)   Each consumer's preferences are described by a utility function with positive first and negative second derivatives.
(ii) Each producer's set of technical possibilities are described by a production function with positive first and negative second derivatives.
(iii) Competitive behavior assumes that the quantities demanded and supplied will be equaled in every market, and that excessive profits will be eliminated.
(iv) Marginal utility and marginal cost determine equilibrium in the market, and marginal productivity and marginal disutility determine equilibrium in the factor market.
(v)   There is perfect competition.

Given the above, I shall now formulate the decision schedule of the orthodox neoclassical agent. The agent is assigned a utility function $u = f(x_1, x_2, \ldots, x_n)$ and a budget constraint that limits his stock of purchases. This budget constraint is expressed as

$$\sum_{i=1}^{n} p_i x_i = y$$

where p_i represents the price of the ith commodity and y designates the total expenditures. The following represents the agent's decision schedule according to the principle of utility maximization:

(i) We form the following Lagrangian function:
$L = u - \lambda (\sum_{i=1}^{n} p_i x_i = y)$ where λ is an undetermined Lagrangian multiplier.

(ii) Assume $\sum_{i=1}^{n} p_i x_i = y$ and obtain $n + 1$ first order conditions by differentiating (i) and setting all derivatives equal to zero. We obtain $\frac{\partial u}{\partial x_i} = \lambda p_i$ ($i = 1, 2, \ldots, k$).

(iii) From (ii) we can obtain $\frac{\partial u / \partial x_i}{\partial u / \partial x_j} = \frac{p_i}{p_j}$. This states that at equilibrium the ratios between marginal utilities and their corresponding prices are equal.

(iv) Second order conditions are required to ensure optimization. This is obtained by requiring that the relevant bordered Hessian determinants alternate in sign.

It is useful to reiterate at this point that the agent's utility function is strictly quasi-concave and that indifference curves are strictly convex. Note further that the consumer has no bliss point and that the utility maximizing commodity bundle will contain positive elements of all commodities.

Demand Functions

To determine the consumer's demand function, one turns to the $(n + 1)$ first order conditions and solve for x_i ($i = 1, \ldots, n$). These solution values represent the amounts of commodities purchased by the consumer. We obtain $x_i = \frac{y}{2p_i}$ for ($i = 1, 2, \ldots, n$). A useful derived result from analysis of the consumer's demand functions is that the consumer's demand functions are homogeneous of degree zero in prices and income.

The Slutsky Equation and Consumer Choice

From the above formulations concerning the consumer's utility and demand functions, the theorist is able to derive a most important result predicted for the consumer's choices when prices and income change. In the specific case of a price change for some item purchased by the consumer, the impact of this is separable into income and substitution effects. This result is known as the Slutsky equation, a central theorem of neoclassical economic theory. This equation is derivable from the *total* differentiation of the above stated first order optimization equation. It is expressible as:

$$\frac{\partial x_i}{\partial p_j} = \frac{\partial x_i}{\partial p_j}\bigg|_{u=\text{const.}} - x_j \frac{\partial x_i}{\partial y}$$

In the above, $\frac{\partial x_i}{\partial p_i}$ and $-x_j \frac{\partial x_i}{\partial y}$ represent the substitution and income effects respectively. According to the theory, it is also argued that the own substitution effect is always negative according to the second order conditions for a constrained maximum. This principle does not apply, however, in the case where there are the changes in the price of one commodity and the quantity demanded of another. Note, too, that it is in the context of the Slutsky equation that results concerning substitute and complement items are generated.

Of course, the central methodological issue here is that concerning the law of demand. The neoclassical theory states that the law of demand is negatively sloped and that this is supported by empirical evidence. This may be the case for large-scale market studies, but the law is somewhat problematic for individual demand schedules. Consider the anomalous demand schedules for Giffen goods and prestige items. Thus, granted the great range in income and general disposition of individuals, it should be obvious that agent demand theory applies only to the prescriptive choice patterns of the ideal neoclassical agent.[8]

The Optimal Firm

Assumptions similar to those for consumer choice optimization are employed for the neoclassical firm's optimal operation. Instead of the consumer's utility function, there is the firm's production function—$F(q_i,\ldots q_n, x_1,\ldots x_n) = 0$—which possesses the following characteristics: (1) the production function is defined only for nonnegative values of the input and output levels; (2) the production function is an increasing function of $q_1,\ldots q_n$ and a decreasing function of $x_1,\ldots x_n$; (3) the production function possesses continuous first and second order partial derivatives; (4) the production function is strictly quasi-concave for output maximization and cost minimization; and (5) the production function is strictly concave for profit maximization.

The neoclassical firm also possesses (a) a family of isoquants (the consumer analog is a family of indifference curves) descriptive of substitutable input factors that yield equal amounts of output, and (b) these isoquants are assumed to be strictly convex to the origin of the production frontier map.

The operational goal of the optimal neoclassical firm is profit maximization subject to relevant information concerning constrained output maximization and cost minimization. Profit is defined as follows:

$$\pi = \sum_{i=1}^{n} p_i q_i - \sum_{j=1}^{s} r_j x_j$$

The optimal firm maximizes profit according to the following rule:

$$J = \sum_{i=1}^{n} p_i q_i - \sum_{i=1}^{s} r_j x_j + \lambda F(q_1, \ldots, x_n).$$

Standard first order differentiation yields the following:

$$\frac{\partial J}{\partial q_i} = p_i + \lambda F_i = 0, \ i=1,\ldots,n.$$

$$\frac{\partial J}{\partial x_j} = -r_j + \lambda F_{s+j} = 0, \ j=1,\ldots,s.$$

$$\frac{\partial J}{\partial \lambda} = F(q_1,\ldots,x_n) = 0.$$

From the above, one can derive the following:

(i) $\dfrac{P_j}{P_k} = \dfrac{F_j}{F_k} = -\dfrac{\partial q_k}{\partial q_j} \quad j, k=1,\ldots,s.$

This means that the RPT (rate of product transformation) for every pair of outputs equals the ratio of their prices:

(ii) $\dfrac{r_j}{r_k} = \dfrac{-\partial x_k}{\partial x_j} \quad j, k=1,\ldots,n.$

This means that the RTS (rate of technical substitution) for every pair of inputs must equal the ratio of their prices.

As in the case of consumer utility optimization, second order conditions for profit maximization are determined by the assumption that the relevant bordered Hessian determinants alternate in sign.

One should note that the optimal neoclassical firm does not possess operational results similar to those of the consumer for the case of input price changes. The change in price for any input yields only a symmetric substitution effect as the total effect of a price change. Recall that the neoclassical consumer's response to a price change in his commodity set is expressible as both income and substitution effects.

In the above, I have examined the operational structures of contemporary neoclassical theory. It is evident that there are important questions that must be asked from the standpoint of the methodology of science. In the first place, the number of restrictions placed on agent choice in the form of the neoclassical theory's assumptions would be of operational significance if the theory's predictive results were borne out by fact. Consider, for example, the restrictions placed on agent choice in terms of the requirements that the consumer or entrepreneur possesses a utility function and a family of strictly convex indifference curves or isoquants. Yet, despite the fact that some agent might truly desire to maximize utility, human error, based on lack of knowledge or invalid inference, could lead to the nonpossession of strictly convex indifference curves. Consider the important Slutsky equation as an example of the indeterminate relationship between the theory of agent choice and its derived laws and theorems. Recall that the Slutsky equation is decomposable into a substitution effect and an income effect. The substitution effect is always negative but the sign of the income effect is indeterminate. In the case of inferior goods, the income effect is positive while for normal or superior goods, the income effect is negative. Note that a refutation of the idea of the negatively sloped curve is evident in the case of the negative income effect of an inferior good. Yet the theoretical model of agent choice possesses only downward sloping demand curves as a test implicational statement. The theory is able to justify empirical evidence for the downward sloping demand curve only "by the addition of an extra auxiliary assumption, asserting the likelihood that any positive income effect will be too small to offset the negative substitution effect of a price change."[9]

Yet, the neoclassical theorist believes that despite evident methodological shortcomings, neoclassical theory constitutes a scientific research program. For example, Robert Russell and Maurice Wilkinson, two textbook authors, argue that it is "comparative—static analysis which yields the predictive, or positive content of the modern theory of the consumer."[10] But the same authors write:

It would appear that the general theory of the consumer does not have a great deal of empirical content. Because of the ambiguity of the income effect, it is simply not possible to predict the direction of change in demand when price changes without making more specific assumptions about the consumer's preferences.[11]

In light of the above discussion, can one argue that given the axiomatic assumptions of neoclassical theory descriptive of the idealized behavior of rational economic man, it follows that logically derived laws such as the law of demand are best construed as purely theoretical propositions? For if the axiomatic assumptions of neoclassical theory are not founded on empirical data (direct or indirect), then propositions derived from them cannot offer acceptable descriptions of empirical events. It is for this reason that theorists like Mark Blaug could argue that "the theory of consumer behavior remains an *ex post facto* rationalization of all final demand outcomes, whatever they are. We can never disconfirm it."[12]

Having set down the essential aspects of contemporary neoclassical ordinal theory, both axiomatic and operational, one sees immediately that there are important methodological issues not yet resolved. To complete the picture, I should want now to examine the methodological discussions that accompanied the development of contemporary neoclassical theory and its ordinalist-choice theoretical base.

Methodology and Ordinal Theory

Contemporary discussion on the methodology of neoclassical theory cannot avoid making references to the ideas of Milton Friedman and Paul Samuelson. Friedman's ideas on methodology are expressed in "Methodology of Positive Economics"[13] while Samuelson's "Problems of Methodology—A Discussion"[14] offers useful insights to his views thereon. Friedman's approach, described by some as instrumentalism, attempts to judge an economic theory in terms of its predictive adequacy. Implicit in this thesis is the notion that the truth content of a theory is of less importance than its predictive capacity. According to Friedman:

Viewed as a body of substantive hypotheses theory is to be judged by its predictive power for the class of phenomena which it is intended to "explain." Only actual evidence can show whether it is "right" or "wrong" or better tentatively "accepted" as valid or "rejected." As I shall argue at greater length below, the only relevant test of the *validity* of a hypothesis is comparison of its predictions with experience.[15]

A ready answer to this position is that since orthodox scientific theory requires that a genuine scientific theory make not only predictions but also offer cognitively satisfying explanations, it would appear that the realism of the fundamental assumptions of a theory is just as important as its predictions. There have been a number of responses to Friedman's thesis notably those of Koopmans,[16] Rotwein,[17] Simon,[18] and Samuelson.[19] It will be more useful, however, to examine the ideas put forward by Samuelson.

Samuelson's response to Friedman's instrumentalist methodology (to which Samuelson refers as the F-twist) is that epistemological claims about the consequences or implications of a theory apply equally to the assumptions of the theory as to the whole theory itself. The basis of Samuelson's thesis is his belief that a genuine scientific theory must be descriptive of the empirical world in its relevant dimensions.

Consider the following:

Define a "theory" (call it B) as a set of axioms, postulates, or hypotheses that stipulate something about observable reality. (If no conceivable observation can even in principle refute, confirm, or touch or bear upon the axiom system taken as a whole, then B is not economics, astronomy, physics, biology or anything properly called science. It might be a model of language, mathematics, mathematical probability or geometry, or game playing—but that is something different.)[20]

This leads to Samuelson's major theoretical claim that the assumptions of a genuine scientific theory, say, (A), are logically equivalent to the theory itself (B), which in turn is logically equivalent to the consequences (C) of that theory. But recall that Samuelson's methodology requires that a scientific theory must be grounded in empirical reality. Thus, if $A = B = C$ and B is empirically grounded, then A and C must also be empirically grounded. This is the basis for Samuelson's descriptivist methodology. Samuelson further demonstrates his commitment to a strict empiricist methodology when he claims that revealed preference theory is a definite methodological improvement on the orthodox ordinal theory. Samuelson writes:

The doctrines of revealed preference provide the most literal example of a theory that has been stripped down to its bare implications for empirical realism. Occam's razor has cut away every zipper, collar, shift and fig leaf. In 1938, I had shown that the regular theory of utility maximization implied, for the two good cases, no more and no less than that "no two observed points on the demand functions should ever reveal the following contradiction of the Weak Axiom":

$$P_1^a Q_1^a + P_2^a Q_2^a > P_1^a Q_1^b + P_2^a Q_2^b$$
$$P_1^b Q_1^b + P_2^b Q_2^b > P_1^b Q_1^a + P_2^b Q_2^a [21]$$

The claim Samuelson makes is that revealed preference theory as an empirical theory is logically equivalent to the pure ordinal theory. This view is reinforced when Samuelson makes reference to the Houthakker's development of the weak axiom to yield the strong axiom of revealed preference, then goes on to argue that the strong axiom of revealed preference is in fact equivalent to orthodox ordinal theory.[22]

Before continuing the discussion on the methodological approaches of both Samuelson and Friedman, one should recognize that their analyses indicate the problems involved in attempting to formulate an adequate methodology for a scientific economics. Friedman's approach in particular testifies to the great difficulty that the economist as methodologist experiences in attempting to formulate empirically grounded assumptions and laws. One notes that much of the history of methodology has been preoccupied with establishing measures for determining the empirical content of the assumptions of a theory. This was the basis for the original formulation of the cardinal utility theory and its ultimate rejection.

Acknowledging the difficulties involved in demonstrating empirical warrant for the assumptions of neoclassical theory, Friedman attempts to show that similar considerations apply in genuine scientific theory. Analogously, a similar approach is urged for economic theory. Friedman's thesis is based, in part, on the argument that the law of falling bodies $s = 1/2gt^2$ is used to predict the fall of bodies in conditions that may not be equivalent to those conditions for which the law ideally applies, that is, free fall in a vacuum. According to Friedman:

> The application of this formula [$s = 1/2gt^2$, where s is the distance traveled in feet and t is time in seconds] to a compact ball dropped from the roof of a building is equivalent to saying that the ball so dropped behaves as if it were falling in a vacuum.[23]

Likewise, Friedman argues that:

> It is only a short step from these examples to the economic hypothesis that under a wide range of circumstances individual firms behave *as if*

they were seeking rationally to maximize their expected returns (generally if misleadingly called "profits") and had full knowledge of the data needed to succeed in this attempt; *as if*, that is, they knew the relevant cost and demand functions, calculated marginal cost and marginal revenue from all actions open to them, and pushed each line of action to the point at which the relevant marginal cost and marginal revenue were equal.[24]

But Friedman fails to recognize that in the case of explanations of the fall of objects based on the ideal case of the free fall of objects in a vacuum, empirically confirmable experiments involving the free fall of objects have been performed in vacuum conditions. On the other hand, the question of what constitutes "rational" behavior is still debated by methodologists of science.[25]

There are further grounds for concern about Friedman's instrumentalism, namely, that to accept any theory that yields accurate predictions as a genuine scientific theory is to create conditions for the accepting of any theory as scientific as long as it yields accurate predictions. Such theories abound, for example, in pseudosciences like, say, astrology. But what demarcates a genuine scientific theory from one that is not is that the assumptions of that theory are stated in terms that could be empirically confirmed. Thus according to the dictates of scientific methodology, a set of assumptions A is sufficient for some theory B if and only if A happens to possess certifiable empirical content.[26]

A useful critique of Friedman's instrumentalism is that of theorist of science Ernest Nagel.[27] Nagel argues that Friedman's thesis is somewhat unclear about the role of certain kinds of theoretical terms, that is, ideal type terms, in the testing of theories. He claims instead that theoretical terms (describing nonempirical pure cases) are of specific usefulness in theory formulation for purposes of empirical testing. But according to Nagel, Friedman would want to dispense with such terms as being "scientifically otiose." There are problems with Nagel's assumption, however, that ideal type terms are structurally cognate in both the natural and social sciences. Consider his claim that the

> Law of the lever in physics is stated in terms of the behavior of absolutely rigid rods turning without friction about dimensionless points; similarly a law of pricing in economics is formulated in terms of the exchange of perfectly divisible and homogeneous commodities under conditions of perfect competition.[28]

But such terms when used in neoclassical theory derive from agent behavior founded on the problematic concept of rational behavior.

Note, too, that while measurement is of extreme importance in ideal type formulation in the physical sciences, such is not the case for ideal type constructs in neoclassical economic theory founded as they are on ordinal rankings rather than on strict measurements. But Nagel does make the important observation that it is unclear in Friedman's thesis whether or not scientific theories ought to explain as well as predict.[29] Quite obviously, the evaluation of the worth of a theory purely on its predictive success without concern as to the reasons thereof is to ignore one of the important criteria required for genuine scientific theory formulation.

Similar reservations about Friedman's instrumentalist approach to economic theory are expressed by A. Coddington,[30] who argues that if it is implicitly understood by Friedman's thesis that the predictive capacity of a theory subsumes its explanatory power, then Friedman's methodology is in error because explanation and prediction are not logically equivalent. One should also remark on Coddington's claim that Friedman's thesis, which states that there is no fundamental methodological difference between social and natural science, fails to recognize that while in natural science research empirical observations require only a "third-person viewpoint," this is not the case in the social sciences where the first-person viewpoint of the social agent could be ontologically incompatible with the third-person viewpoint of the observer. In fact, it is just this incompatibility that engenders questions about the nature of "rational behavior."

Friedman's thesis, however, is favorably reviewed by theorists such as Boland[31] who argue that "Friedman's essay [on methodology] is an instrumentalist defense of instrumentalism" (Boland, 1979, p. 522) and that "The repeated attempts to refute Friedman's methodology have failed, I think, because instrumentalism is its own defense and its only defense" (p. 522). Consider, too, Boland's comments on instrumentalism: "Finally, and most importantly, I think it essential to realize that instrumentalism is solely concerned with (immediate) practical success" (p. 520). In this regard, a ready test for Friedman's instrumentalist approach would be to determine how successful are his theories concerning economic behavior. Thus, one might want to consider Friedman's espousal of the Quantity Theory of Money in Monetary Economics. According to Friedman's thesis, the success of this theory should be determined by its predictive successes. But the Quantity Theory has not been shown to be a successful predictor of behavior of the economy as a whole. There is no unanimous agreement within the economics profession that Friedman's Quantity Theory is valid. Its predictive failures militate against this.

But quite obviously, an examination of the theory's assumptions would be one way in which the causes of its predictive failures could be determined. The empirical content of a theory's asssumptions is crucial, therefore, in determining its scientific credentials.

Yet Samuelson's descriptivist approach to economic theory is regarded as being equally problematic by some theorists. Machlup, for instance, argues that Samuelson's methodology as expressed in "Problems of Methodology" leads to the rejection of all theory. According to Machlup, "A theory, by definition, is much wider than any of the consequences deduced,"[32] and that:

> We never deduce a consequence from a theory alone. We always combine the postulated relationships (which constitute the theory) with an assumption of some change or event and then we deduce the consequence of the conjuction of the theoretical relationships and the assumed occurrence.[33]

Machlup goes on to argue, with reference to Samuelson's papers on "international factor-price equalizaton," that Samuelson

> sometimes uses language of empirical operations, for example, when he speaks of "observing the behavior of a representative firm." It should be clear, however, that what he "observes" is merely the logical consequence of a set of assumptions; that the "behavior" is purely fictitious; and that his representative firm is only an ideal type, a theoretical construct.[34]

Finally, Machlup states that

> Samuelson, one of the most brilliant theorists in present-day economics, produces his best work when he deduces from unrealistic assumptions general theoretical propositions which help us interpret some of the empirical observations of the complex situations with which economic life confronts us.[35]

It would seem from the above discussion that Machlup's views on methodology are more in accord with Friedman's views than with those of Samuelson. Samuelson responds, however, by making the claim that it is better that the assumptions of a theory be empirically confirmed than not.[36] In other words, Samuelson is asking that the factual content of the assumptions of a theory be of decisive importance in the determination of whether to reject it in favor of a rival theory, or to seek its modification. According to Samuelson, it is the level of discrepancy between empirical fact and the implications of the assumptions of some theory that determine whether or not its theorists should retain those assumptions.

Samuelson's emphasis on the importance of a theory's assumptions in determining its tenor no doubt underlies his insistence that scientific theories describe events and processes in the world rather than explain them—in the sense of answering ultimate "why" questions. Thus, for Samuelson, explanation is nothing more than a synonym for description. He would claim that "Scientists never explain any behavior, by theory or by any other hook. Every description that is superseded by a 'deeper explanation' turns out upon careful examination to have been replaced by still another description."[37] A scientific explanation, for Samuelson, is nothing more than "a superior description in that it successfully fits a wide range of empirical regularities."[38]

But Samuelson does not recognize that there are good semantical reasons why the term "explanation" is used instead of "description" for purposes of appraising scientific theories. Given that in complex scientific theories where the facts under analysis are not empirically self-evident—but are linked together by means of inference and hypothesis—it is possible that different theories could seek to account for a given set of results. It is easier to imagine disagreements in explanations of unemployment than about factual claims that "the prime rate of interest is twelve percent." Samuelson's descriptivist thesis fails to assign sufficient importance to theoretical terms in theory formulation. It is true that some term that is now regarded as a theoretical term in a given theory may later be regarded as an observational term because of the use of improved instruments, and it is on these grounds that successful theories need not be regarded as invalid when there is no direct empirical proof for the empirical content of their theoretical terms. Yet these considerations cannot be truly addressed by Samuelson's thesis given its requirement that a scientific analysis be essentially a set of descriptive statements.

The discussion and analysis of the methodological views of Friedman and Samuelson demonstrate that the ideas on methodology of these two important theorists may be regarded as representative of a diametrically opposed instrumentalism and descriptivism. It appears, though, that most orthodox contemporary theorists support a model that is explanation-seeking while displaying features of instrumentalism and realism, as in the case of Wong, or tend toward supporting some variant of instrumentalism, as in the case of Boland.

Wong, for example, argues against Samuelson's descriptivism by claiming that a purely descriptivist methodology does not fully represent the structure and role of scientific theories. We recall that Samuelson argues that scientific explanation is ultimately synonymous with description. Wong supports his thesis by claiming the following:

> According to Samuelson's descriptivist methodology, (1) a theory is just a description of observable experience, a convenient and economic representation of empirical reality (1952, p. 61); (1963, p. 236); (1965a, p. 1171). As a corollary, (2) knowledge consists essentially of observational reports, a theory expressible in observational language is superior to those that are not. The reasons for rejecting the view that theories are explanatory are: (3) Explanations are ultimate (1965a, pp. 102–03). (4) Apriorism must be avoided, hence theories should be expressed in observational language (1963, p. 235); (1964, p. 738). (5) Explanations turn out to be just better descriptions (1964, p. 737); (1965a, p. 1165). (6) All well-known theories in science are expressible in terms of observational statements, i.e., basic statements (1965a, p. 1167).[39]

Wong's answer to Samuelson's descriptivism is to stress the role that explanation plays in a genuine scientific theory. According to this theorist, the model that appropriately fits economic theory is the deductive nomological model of scientific explanation. Recall that in this model the explanans logically entails the explanandum, but the converse is not the case. It is also required that the explanans contain at least one unrestricted universal statement. And given that "an unrestricted universal statement is not equivalent to a finite conjunction of observational statements, i.e., basic statements" (Wong, p. 320), then it is evident that "a theory is not logically equivalent to a set of observational statements because of the logical form of a theory" (p. 320).

Wong's analysis of Samuelson's descriptivism is useful in that it demonstrates the epistemological weaknesses of a purely descriptivist methodology, but his appeal to the orthodox idea of a scientific theory to support his claims is not really relevant to the structure of ordinal microeconomic theory. For there is no evidence that orthodox neoclassical theory possesses at least one unrestricted universal statement, which according to Wong, is a necessary requirement for a scientific theory. It should be emphasized that the postulate of rationality serves as a key initial condition in neoclassical theory, but this postulate and its derived axioms and theorems are to be properly regarded as prescriptive since they cannot be shown to correspond in a cognitive way to the actual choices of economic agents. And more importantly, epistemologists have not yet been able to agree on what "rationality" really means.

The point discussed is important since it could be argued that Samuelson's and Friedman's methodologies deemphasized the role of explanation in economic analysis because of the problematic nature of the concept of rationality. One should note that the concept of explanation in science, which entails answers to "why" questions, does not allow similar treatment for both the natural and social sciences. In the case of the natural sciences, explanation affords cognitive satisfaction when empirical evidence whether direct or indirect is forthcoming and publicly confirmable. But explanation of agent behavior in the social sciences, and specifically in the case of economics requires the formulation of explanans statements in terms of empirically inaccessible private mental states, all theoretically reducible to the concept of rational behavior. In this sense, the deductive nomological model of scientific explanation is not compatible with the basic assumptions of the orthodox neoclassical theory of agent choice.

Wong's useful critique is deficient in the sense that it does not discuss much the foundational postulates of economic theory. Yet it is analysis of the foundational postulates of economic theory that affords the possibility of adequate explanation. But as is evident, the formulation of an adequate model of explanation in economic theory is highly problematic. The solution, however, is not the adoption of an instrumentalist program. It is generally accepted that a viable scientific theory should not only make predictions but must also offer explanations. In fact, "explanation" appears to be more of a necessary criterion of theoretical science than prediction—because it is "explanation" that offers the possibility of understanding the phenomenon or event under analysis. Furthermore an understanding of the phenomenon under analysis requires that the assumptions of the theory be analyzable in empirical terms, direct or indirect.

In fact, there is much evidence to support this view from the study of the history of science. Consider that traditional taxonomic biology was forced to give way to molecular biology and biochemistry because of the increased explanatory power offered by these newer disciplines. And these disciplines offer explanation not at the macroscopic level but increasingly at the microscopic level. For example, modern genetics does not offer explanations in terms of phenotypical classification but in terms of microscopic analysis of the chromosome.

What is suggested here is that to focus exclusively on the

predictions of a theory without regard for the importance of the empirical content of its assumptions does not allow the possibility for reductionism in the case of economic theory. Despite misgivings, it may be that psychology offers better answers to the puzzles of human decision making than those offered by the axiomatic assumptions of microeconomic theory. But again, this brings us to the well-known difficulty of analyzing mental states, which are at the base of agent decision and choice.

It is also useful to point out that some authors argue that an acceptance of instrumentalism entails the thesis that it is futile to criticize the neoclassical maximization hypothesis. Lawrence Boland, for example, writes:

> The truth of assumptions supposedly matters to those economists who reject Friedman's instrumentalism, but for those economists interested in developing economic theory for its own sake, I have argued here that it is still futile to criticize the maximization hypothesis.[40]

Given that the neoclassical maximization hypothesis is a central postulate in orthodox neoclassical theory, it is evident that Boland's argument is an attempt to insulate the instrumentalist defense of neoclassical theory from criticism. I will show, though, that this task is unsuccessful. Boland's thesis rests on the claim that the "actual form or the neoclassical premise is not a strictly universal statement."[41] The neoclassical premise, according to Boland, is an "All and some statement"[42] properly expressible as "for all decision-makers there is something they maximize."[43] Furthermore, all and some statements are neither refutable nor verifiable. They are not verifiable because universal hypotheses (All x have property y), according to inductive logic, are not verifiable (though justifiable); nor can existential statements (some x have property y) be refutable (even if false).

Boland's proof is that any claim based on empirical evidence that purports to show that some agent x has not maximized anything can be responded to by stating that it is logically impossible for the claimant to have exhaustive proof that there is nothing that the agent is maximizing. According to Boland:

> The verification of the counter-example requires the refutation of a strictly existential statement; and as stated above, we all agree that one cannot refute existential statements.... In summary, empirical arguments such as Simon's or Leibenstein's that deny the truth of the maximization process are no more testable than the hypothesis itself.[44]

Boland claims finally that the maximization hypothesis is not a tautology but a metaphysical statement that serves a methodological function in neoclassical theory. Thus, any critique of the maximization hypothesis should be addressed not to the truth of that assumption but to neoclassical methodology as a whole. Boland supports this last claim by stating that every scientific research program has at its core a set of basic nontestable metaphysical propositions. So "The neoclassical assumption of universal maximization could very well be false, but as a matter of logic we cannot expect ever to be able to prove that it is."[45]

Boland's argument cannot withstand serious scrutiny since its claim that the maximization hypothesis is an "All and some statement" is subject to question. First of all, he seems to have misinterpreted seriously the form of the neoclassical maximization hypothesis. In fact, the neoclassical maximization hypothesis is not really expressible as an "All and some statement." Recall that the initial assumption of the neoclassical theory is that the neoclassical universe is populated with rational agents whose choice paths are determined *a priori* by the axiom set of the theory. Note that the axioms of completeness, reflexivity, and transitivity are among the most important of the neoclassical axiom set. Furthermore, the neoclassical theory is quite specific about the "something" that agents are required to maximize. The proper form of the neoclassical maximization hypothesis is, therefore, "all neoclassical consumers maximize utility, and all neoclassical entrepreneurs maximize profits."

More specifically, the normal maximization process for the consumer is expressed as follows: $U = f(q_1, q_2 \ldots q_n)$, which in turn is subject to the budget constraint

$$y - \sum_{i=1}^{n} p_i q_i = 0$$

In fact, one of the exercises that students of orthodox neoclassical microeconomics are trained to perform is the determination of what consumption bundle the rational consumer should purchase in order to maximize utility. The consumer who does not purchase the recommended consumption is regarded as irrational by definition. In fact, proof that the neoclassical maximization hypothesis is not an "All and some statement" is that research results demonstrate that agents do make "irrational" choices.[46] Quite obviously, the maximization hypothesis is not an "All and some statement" but rather a universal statement. This universal statement is indeed falsifiable and has been falsified if one consults the empirical evidence. Clearly, if Boland's hypothesis were correct, then there would be no empirical evidence of irrational agents since according to his formulation of the maximization hypothesis "for all decision makers there is something they maximize." Yet a puzzling question remains: since the neoclassical maximization is founded on a set of axioms that define rational behavior, is Boland arguing that all agents are responsible for their own versions of rationality? According to the neoclassical theory, however, this would be unacceptable since the theory postulates only one version of rationality.

Yet, if one accepts the notion that the neoclassical agent is a rational decision maker, then there is a sense in which the maximization hypothesis is neither confirmable nor falsifiable in the scientific sense of those terms. It is generally believed that the proposition "All good men pay taxes" is a normative proposition because of the noncognitive, emotive content of the term "good." But I believe that the same holds for the proposition "All rational men pay taxes," since the terms "good" and "rational" signify a willed disposition to make certain choices. Not all men are disposed to be "good" or to be "rational" according to prescribed criteria. I will pursue this discussion in the following chapter.

To return to Boland's thesis that "the neoclassical assumption of universal maximization could very well be false, but as a matter of logic we cannot expect ever to be able to prove that it is,"[47] I argue that from the context of neoclassical theory that the maximization hypothesis is indeed falsifiable, and that empirical results have shown that the neoclassical maximization hypothesis is unreliable in its capacity to predict and explain decision making. Instrumentalism as a justificatory methodology is of little value for theories weak in predictive capacity.

Expected Utility Theory

In the above sections, I have discussed ordinal utility theory and its role in shaping the structure of orthodox neoclassical theory. This theory traditionally focuses on the strictly deterministic choices of the individual according to the fundamental axioms of choice. But ordinal deterministic utility theory could be viewed as complementary to another form of utility-based decision making under conditions of risk, that is, expected utility theory. This theory encompasses decision making not only in economics but also in operations research and psychology. Although the theory of expected utility has its basis in the eighteenth-century formalizations of mathematicians like Daniel Bernouilli (noted for his attempts to explain the St. Petersburg paradox), modern efforts are due to von Neumann and Morgenstern, and Jacob Marschak. The theory of expected utility developed by these theorists demonstrates that the rather complex mathematical computations that the individual must make in order to conform to the expected utility model are often not made. The research of Tversky and Kahneman offer proof of this, and reinforce further the normative structure of general ordinal utility theory. The five axioms[48] of choice originally introduced by von Neumann and Morgenstern, though based on probability estimates, play the same role in expected utility theory as the axioms of choice in deterministic ordinal theory. And according to the theory, it is to be understood that the rational decision maker would conform to these axioms.

A Summary and Postscript on Methodology

Granted the misgivings that neoclassical theorists of economics have developed concerning the research methodology introduced by Friedman and Samuelson, the attention has shifted to the methodologies of scientific research proposed by Popper and Lakatos. For example, Blaug's criticisms of current neoclassical theory derive in the main from an appeal to Popper's falsifiability criterion. Blaug believes that one way in which neoclassical economists could improve the scientific image of their discipline would be to seek to introduce more falsifiable hypotheses into their research efforts.[49] But Popper's stern methodology of research with its emphasis on theory rejection based on falsifiability seemed to threaten the neoclassical program with a fate similar to that reserved by Popper for Marxism and psychoanalysis. Thus, Popper's research methodology was not particularly welcomed by the neoclassical theorist.[50]

But the less stringent and more accommodating appraisal of scientific methodology on the part of Lakatos seems to be attracting a larger audience.[51] I believe that the attractive feature of Lakatos's methodology as far as neoclassical theorists are concerned is the role that concepts such as "irrefutable hard core," "negative heuristic," "positive heuristic," "protective belt," and so on play in scientific research programs. No doubt, given the problematic nature of neoclassical theory, Lakatos's methodology is of heuristic value since it would appear to offer justification for the neoclassical theorist's defense of his discipline by retreating behind its "irrefutable hard core" and "protective belt." There is scientific respectability in taking this route, since Lakatos's methodology was developed with physical science in mind, the neoclassical theorist's model science.

But note first of all that Lakatos expresses little concern for the methodology of the social sciences. His reflections on matters such as progressive or degenerating research programs applied only to the physical sciences—disciplines long recognized as genuine empirical sciences. Yet this does not deny that economics, defined by Kuhn, Lakatos, and others as an immature scientific discipline, could mature into a genuine scientific discipline. However, this possibility could not be fully explored without considerations given to what constitute adequate criteria for determining a scientific discipline.

In the postpositivist discussions on the methodology of scientific research, Popper has gained the reputation of being the most stringent and critical in terms of criteria necessary for determining the scientific credentials of a theory. The purpose behind Popper's stringency, as pointed out, was to discredit the claims to scientific status made by research schools such as Marxism and psychoanalysis. Kuhn's history of scientific progress, on the other hand, argued for epistemological relativism as the dominant feature of the history of science. The narrowing of the epistemological division between natural science and other fields of inquiry, suggested by Kuhn's approach, attracted the social scientists despite his claims that the latter were not yet developed sciences. However, Kuhn's research efforts were mainly descriptive of what he viewed as the history of science; he offered little in terms of the methodology of scientific research. It is Lakatos's relatively looser and more flexible criteria that seem to offer the neoclassicist economist the needed methodology to salvage his discipline once again.

But consider the methodological caveat coming from the unlikely source of Feyerabend, an anarchist in the area of epistemology. Feyerabend argues that Lakatos's methodological standards are quite loose, as the following demonstrates: "Scientific method, as softened up by Lakatos, is but an ornament which makes us forget that a position of 'anything goes' has in fact been adopted."[52] On this basis, Lakatos's methodological standards would admit research-program status not only to economics but also to astrology. But this is hardly what the neoclassical theorist wants. He wants physics and neoclassical economics in the same ontological framework, not astrology and neoclassical economics.

The neoclassical theorist would, no doubt, maintain confidence in the Lakatosian methodology and argue that neoclassical economics is a budding research program and needs time to develop. But this thesis is seriously questioned by Alexander Rosenberg, who takes issues with E. Roy Weintraub's recent attempts to defend neoclassical general equilibrium theory on the grounds of it being a "budding scientific research program."[53] It is useful to point out that Rosenberg is, perhaps, one of the few theorists who has recently reserved strong criticism for the scientific pretensions of neoclassical theory.[54]

Yet the skeptical analyses of contemporary neoclassical theory continues, though always with the prescription that with the appropriate methodological changes here and there, the discipline could evolve into a science. The most recent effort in this context following the methodological efforts of Blaug, Boland, and Caldwell is that of Donald McCloskey.[55] In an interesting text, McCloskey asks for an improvement in the art of doing economics by recommending that economists pay more attention to their modes of communication, that is, their rhetoric. According to McCloskey, the neoclassical economist's normal method of communicating was mired in a constricting scienticism that hardly did justice to the complexity and variegated nature of economic decision making. McCloskey's thesis is that to study the rhetoric of economics is to explore economics in its fullest dimensions as a *science humaine*. His justification for this prescription is that he views economics not as a predictive but as a historical science.

The significance of McCloskey's work derives, I believe, from the attention it brings to the "modernist" language of neoclassical economics and its crippling effects on the thinking processes of economists. Yet I want to argue that McCloskey's claim that the methodology of scientific investigation should be given up in favor of the rhetorical or conversational model cannot be sustained. He supports this thesis with the argument that the idea of "methodology" has fallen out of favor in some philosophical quarters. But this point of view belies the fact that empirical science (physics, molecular biology, etc.) has relied on and continues to rely on a strict methodology of investigation. This prescriptive methodology is inculcated into the thinking habits of the science student while being trained in laboratory techniques. In fact, the methodology of empirical science research is so strict that individuals who circumvent the rules by doctoring data or employing dubious experimental methods quickly lose their professional reputations within the community of scientists. This is less so in the social sciences.

This observation supports the thesis for which I argue in this text: the physical and biological scientists have worked out for themselves the most efficient and appropriate methodologies for their particular kinds of research. But this methodology is inappropriate for the social sciences in general and economics in particular given the nature of human behavior. The problem with neoclassical economics is that its founding theorists adapted from mechanics a methodology of research that was wholly inappropriate.

I conclude this chapter by pointing out that just as the cardinal utility theory was beset by a number of methodological problems, so, too, is the ordinal theory and the derived structure of contemporary neoclassical theory. In the following chapter, I extend the discussion of contemporary neoclassical theory by examining general equilibrium theory, in order to see that it is affected by methodological problems similar to those of the neoclassical microeconomic theory.

6
General Equilibrium Theory— An Analysis

In the previous chapter, an attempt was made to show that theorists of neoclassical economics have sought throughout the history of their discipline to establish a genuine science of economics sharing a methodology of research similar to that of the natural sciences. But it was also shown that despite the ordinal utility-theory approach introduced by Hicks, Allen, and Samuelson, neoclassical theory has failed to live up to its promise as a scientific discipline. It has failed in the areas of explanation and prediction crucial for the determination of scientific status. The discussion testified to this. This failure was explained by pointing out that the nature of the axiomatic foundations of neoclassical economics (given their reliance on the postulate of rationality) has greatly compromised the scientific pretensions of neoclassical economics. On this account, neoclassical theory reduces to a theory of prescriptive propositions presenting itself as an empirical science.

But the analysis of positive neoclassical economics is not yet complete. In this chapter, I shall examine the structure of the neoclassical general equilibrium theory, the theory of the behavior of all the units of the neoclassical economy. It will be argued that since the axiomatic assumptions (including the important postulate of rationality) of the general equilibrium are identical with those of agent choice theory, any critique of the former in terms of explanation, prediction, and so on should appeal to equivalent considerations raised against agent choice theory. The analysis will conclude with the claim that the axioms, theorems, and derived laws of general equilibrium theory are prescriptive though highly formal propositions lacking in genuine empirical content.

It should be noted first of all that the literature on general equilibrium theory is quite extensive with research that includes the works of Walras, Wald, Arrow, Debreu, Hahn, Scarf, and others.[1] Of interest, too, could be metatheoretical work of such authors like

Handler[2] and Hands[3] who have attempted to formulate that theory in structuralist fashion. I should not be concerned to examine the whole theory though, but would prefer rather to present a brief synopsis of its basic points before engaging in an analysis of its role in the structure of neoclassical economics.

It would be instructive to begin with the usual formalization of the theory. Assume a set of consumers, a set of producers, and states of economy. This theoretical economy contains n goods and services, m consumers, and l producers. The matrix S that equals $[X; Y]$ is such that X represents an $n \times m$ matrix describing the choices of m consumers, and Y represents a matrix of $n \times l$ elements that describes the input-output choices of l producers. It is also assumed that the total consumption of any good is no greater than the total output of that community and that entrepreneurs choose that possible optimal input-output vector which maximizes profits for the firm. The possible inconsistency that arises when the consumer as entrepreneur maximizes profits and appropriates these earnings is averted by defining the entrepreneur as an employee of the firm whose income is equal to the marginal value product of his or her work.

It is important to bear in mind though that implicit in the analysis to follow, the key neoclassical axiomatic assumptions concerning rational agent choice also hold. Reference here is made to the assumptions about (1) completeness, (2) transitivity, (3) non-satiation, (4) substitutability, (5) convexity, and so on.

Having stated the necessary assumptions of neoclassical equilibrium theory, I now sketch the formal operatives necessary to achieve the three essential features of equilibrium theory: its existence, uniqueness, and stability. In standard equilibrium theory, a positive competitive equilibrium price is defined as follows: a positive equilibrium price is a positive price such that the excess demand function at this price is zero. A competitive equilibrium is then defined as a situation in which there is no market excess demand and all existing market prices are nonnegative. More formally, one can state that $\bar{P}_\varepsilon S_n$ is an equilibrium price if $Z(\bar{P}) \leq 0$, where $Z(P)$ represents agent choice behavior. Note that if $P_\varepsilon S_n$, $PZ(P) = 0$ (Walras' law) and that $Z(\bar{P}) \leq 0$, then $z_i(\bar{P}) < 0$ implies that $\bar{P}_i = 0$.

The Existence of Equilibrium

Twentieth century contributions to research on the existence of equilibrium date back to the efforts of Abraham Wald and John von Neumann, but it was the Arrow-Debreu paper of 1954 ("Existence of an Equilibrium for a Competitive Economy") that served as the standard reference for later works on proofs of the existence of equilibrium. It should be noted, though, that Debreu's *Theory of Value* had already established the foundations for subsequent developments. In this brief formulation of the principle of the existence of equilibrium, I prefer to appeal to the more recent analyses of Arrow and Hahn.[4] Accordingly, the following conditions are necessary and sufficient for the existence of a competitive equilibrium.

(1) For any P (price vector), there exists a unique number $z_i(P)$, the excess demand function for i. In general, there exists $Z(P)$ a unique vector of excess-demand functions.
(2) $Z(P) = Z(_kP)$ for all $P > 0$ and $k > 0$. This means that excess demand functions are homogeneous of degree zero in P.
(3) For all $P_\varepsilon S_n$, $PZ(P) = 0$.
(4) $Z(P)$ is continuous over its domain, S_n.

It is useful to point out that Pareto optimality is assumed for any competitive equilibrium and that any Pareto optimal position is a competitive equilibrium. Furthermore, it can be proven that any competitive equilibrium is in the core of the system. The core of the economy is defined as any feasible allocation that cannot be improved on by any coalition of choices.

The Uniqueness of Equilibrium

The notion of the uniqueness of equilibrium implies that there is only one normalized price set and only one state of the economy resulting in market clearance. To establish a proof of this requires an account of the possibility of there being more than one state of the economy associated with an equilibrium set of prices and of the possibility of there being more than one equilibrium price vector. The following assumptions are then necessary.

(a) Assume strictly convex consumption and production sets.
(b) Assume that as the relative price of a commodity increases, the excess demand for it falls—regardless of the values taken on by other prices. From this, it follows that if there is an equilibrium price $P^* > 0$, then it is unique.

Stability of Equilibrium

It has been shown that given certain assumptions about the choice paths of economic agents, there will always exist a price vector that equates supply and demand. But is there a guarantee that the economy will tend to operate at this price vector? In other words, what are the market conditions to ensure the stability of equilibrium? Despite the usefulness of the ideas of Marshall, Walras, and Hicks concerning static stability, contemporary neoclassical equilibrium theory relies on the thesis of the Walrasian tâtonnement adjustment mechanism. This mechanism spells out the manner in which prices and quantities demanded and supplied over time change in response to a disturbance of equilibrium.

Specifically, the Walrasian notion of tâtonnement is founded on the assumption that if the market happens to be at a set of nonequilibrium prices, then at the beginning of each market period a referee announces a price vector that serves as the basis for demand and supply in the economy. If the market is cleared at this price vector, then an equilibrium exists. If not, then the price referee prohibits any trading until supply and demand are equalized by a "step by step" adjustment of prices. At this point, trading is permitted. More formally: (1) If P^* is an equilibrium price vector, then $g_i(z_i(P^*)) = 0$ for all i. (2) In the absence of equilibrium and given an initial set of prices P°, then as t tends to infinity, the price vector P approaches some equilibrium price vector P^*.

To distinguish between the possibility of multiple stable equilibria and a particular equilibrium, assume that $V(P_j^\circ t)$ represents the path of prices P over time. The following results then summarize Walrasian stability:

(1) A system is stable if for any P°, $\lim_{t \to \infty} V(P_j^\circ t) = P^*$ for some P^*
(2) If $\lim V(P_j^\circ t) = P^*$, then there is global stability as $t \to \infty$
(3) Local stability is established if $\lim V(P_j^\circ t) = P^*$ for some neighborhood of P^*.

As a final note on the question of the stability of equilibrium, it should be pointed out that the stability of equilibrium has not been established as a general empirical rule. Some theorists have even pointed out that in theoretical cases of unstable equilibria, the law of demand is inoperative. This observation, no doubt, raises questions about the viability of the tâtonnement process in dynamic equilibrium theory.

General Equilibrium Theory as a Scientific Theory

General equilibrium theory axiomatized is usually regarded as the ultimate formalization of economic theory and is assumed to offer a clear theoretical framework of economic theory. Consider, too, the following observations of Gerard Debreu.

> The benefits of the axiomatization of economic theory have been numerous. Making the assumptions of a theory entirely explicit permits a sounder judgment about the extent to which it applies to a particular situation. Axiomatization may also give ready answers to new questions when a novel interpretation of primitive concepts is discovered.[5]

Yet there is much controversy concerning axiomatized equilibrium theory. Some authors have raised questions about its significance for economic theory. It would appear that much of the controversy on this issue derives from the fact that although general equilibrium theory is deductively valid in a highly formalized way, there is some question as to whether it possesses empirical content, a necessary condition for scientific status. In fact, Debreu himself testifies to this in the following:

> In the past few decades, that wide range of problems has been the subject of an axiomatic analysis in which primitive concepts are chosen, assumptions concerning them are formulated, and *conclusions are derived from those assumptions by means of mathematical reasoning disconnected from any intended interpretation of the primitive concepts.*[6]

But in this context, Hahn, for example, implicitly defends the Arrow-Debreu equilibrium thesis by asserting that the search for equilibrium is still worthwhile despite the fact it may never be attained.[7] In reply to Kaldor's argument[8] that increasing returns to scale militate against Arrow-Debreu types of equilibrium systems, Hahn argues to the contrary and also claims that "it is perfectly possible for an Arrow-Debreu equilibrium to exist even though the axioms of the theory are violated."[9] In the same vein, Hahn argues for the theoretical necessity of general equilibrium theory by pointing out that discussions on economic issues such as exhaustible resources, balanced budgets, and so on are meaningful only within the context of assumptions about Arrow-Debreu–type equilibria. But note, too, the stronger claim by Hahn that the Arrow-Debreu equilibrium can be applied as an empirical tool to falsify propositions and theories in a specific way.[10] Yet the same author writes elsewhere:

> It was, I believe, always understood that the equilibrium of Arrow-Debreu is not a description of an actual economy, and I have already given reasons why the concept should nonetheless be important and interesting. However, one certainly does want a conceptual apparatus which is much more nearly descriptive.[11]

Clearly, Hahn is expressing some doubts about the empirical content—hence scientific credentials about the Arrow-Debreu model. The point is that while formalization is a necessary condition for the construction of a scientific theory, it is not sufficient. It is the specific interpretability of its variables in the empirical sense that gives a formal model its scientific content and allows for potential falsification. And despite strong commitments to formal equilibrium analysis, Hahn is nevertheless constrained to recognize its limitations in the empirical sense. Consider again the following:

> The kind of issues which I have been discussing have been concerned with the conceptual apparatus of economic theory. As such the analysis is almost bound to lack concreteness especially when some of the terrain is so speculative, and this fault causes me no feelings of guilt. But since some of what I have had to say turned on the inadequacy of our present paradigms I fear that the impression may have been gained that I think the latter to be "sterile" and "useless." Nothing could be further from the truth. Not only does the Arrow-Debreu equilibrium continue to be a special ideal type of the notion here proposed, but it is also of great use for many purposes, some of which have already been noted. But the paradigm is of course of ambitious generality and for very many important purposes a much more modest Marshallian apparatus will do very well.[12]

But while Hahn's defense of general equilibrium is somewhat guarded, though optimistic, that of Alan Coddington[13] is much more confident. Coddington argues that Hahn's thesis, which claims that general equilibrium theory should have some empirical implications, is inaccurate. According to Coddington, the relationship between theories and their subject matter should not be judged in terms of any direct relationship between the former and the latter. The evaluation of a theory should not entail, therefore, an appraisal of it in terms of its realism. This would involve what Coddington describes as a "category mistake."[14] After making the assumption that a theory ought to be evaluated in terms of what it sets out to do, Coddington seeks to defend the general equilibrium theory purely in terms of its formal properties, without regard for its falsifiability potential or empirical content.

Coddington's thesis is that the function of the general equilibrium theory is principally formal and that its empirical implications are of minor importance. In response to Hahn's claim that the formal general equilibrium theory not only performs the task of enforcing the precision of ideas pertinent to the theory, but is also instrumental in examining the falsifiability potential of theories, Coddington makes the following counterclaims:

(i) The GE construction does not contribute to the ease with which theories may be falsified (by evidence) but to the ease with which inconsistencies between a theory and a theoretical counter-example, within the formalism, may be established.

(ii) The GE construction does not contribute to the precision with which ideas may be expressed, but rather to syntactical watertightness within a semantically-uncharted sea.

(iii) The General Equilibrium construction does not facilitate the refutation of unsound arguments by the provision of stronger counter-arguments, but rather to the detection of logical defects within unsoundly-formulated arguments.[15]

In sum, Coddington's appraisal of general equilibrium theory is founded on the notion that its status is determinable not by its falsifiable potential but by its logical coherence and its capability to solve problems from within the framework of the theory.[16]

Recall that one of Coddington's main points against Hahn's methodology is that it was too dependent on the Popperian falsifiability thesis: any counterinstance to some theory T is sufficient to refute or falsify T. Coddington's claim against Hahn was founded rather on Lakatos's thesis that falsifying evidence for a theory is usually not viewed as sufficient for a refutation of that theory. He points out that Hahn, though a Popperian in terms of methodology, endorses Lakatos's thesis of the research program in his practice.[17] His aim is to show that anomalous empirical data are not sufficient to overthrow a formal theory such as general equilibrium theory. These are the grounds on which Coddington argues for the theoretical and formal autonomy of general equilibrium theory.

But the basis of this argument cannot be sustained: the fact that the existence of anomalous data is not sufficient to falsify a formalized theory is not necessarily a warrant for the theoretical autonomy of that theory. If the purpose of the theory is to establish the truth or falsify formal propositions, or to demonstrate their logical connectedness by deductive inference, then whether the theory corresponds to events in the empirical world or not is not relevant to the tenor of that theory. Theories in mathematics are judged in this way. If, on the other hand, the theory in question purports to describe or correspond to events in the empirical world, then the sole requirement of "syntactical watertightness" is not sufficient. Economics is a field of inquiry whose purported aim is to examine the utility-laden choices of human agents. On this basis, it has historically aligned itself with physics and chemistry as an empirical discipline, not with pure mathematics or formal logic.

Thus, although a single anomalous event is not sufficient to falsify an empirical theory (*pace* Popper), a series of such anomalous events would tend increasingly to raise questions about the viability of the research program in question. Yet, such a program will nevertheless continue to be sustained until some rival research program with greater explanatory power gains adherents.[18] Coddington's thesis could be sustained if general equilibrium theory were a purely formal theory. But this could not be the case because its axiomatic assumptions make reference to empirical objects. Similar concerns are expressed by such theorists as Blaug[19] and Handler,[20] who express concern about the fact that the general equilibrium theory, supposedly the unified theory of economics, stresses abstract form rather than empirical content. While Blaug believes that this theory is without much scientific significance unless it contains falsifiable theorems, Handler expresses puzzlement at the ontological status of the theory's supposedly referring terms. In short, the general criticism is that the model, constructed by theorists as a formal framework from which meaningful statements about the economy may be made, is too divorced from the empirical events it is supposed to explain and predict.[21] And, of course, the methodological problems that beset the general equilibrium model also affect such derived theories as marginal productivity theory and capital theory.

Given the analysis of general equilibrium theory in the above discussions and my claim that as a scientific theory it is deficient in terms of explanation and prediction, the question then is what function does the theory serve in contemporary neoclassical theory? Alexander Rosenberg argues that economics (this would include neoclassical general equilibrium theory) is "mainly an exercise in applied mathematics."[22] Yet Daniel Hausman argues that although theories of general equilibrium are without explanatory worth, such theories do have cognitive worth. He argues specifically that the existence proofs of general equilibrium theory are of value and that the results of partial equilibrium analysis and practical general equilibrium models offer some hope.[23] Hausman's approach to general equilibrium is thus more exploratory and tentative than otherwise. He entertains the belief that mathematical analysis could explore the possibilities of modifying the axiomatic assumptions of general equilibrium theory.

Consider, too, E. Roy Weintraub's recent attempt to justify general equilibrium theory by appeal to the idea of the scientific research program developed by methodologist of science Imre Lakatos.[24] Weintraub's goal is to offer grounds for the insulation of neoclassical equilibrium theory from criticism on the basis that a research program's hard core may be examined only by rival theorists; its adherents should accept a research program's hard-core propositions without question.[25] Again, Weintraub's commitment to Lakatos's methodology is such that he even claims that the term "theory" is inappropriately employed with regard to general equilibrium theory, preferring instead "model" or "approach" for this branch of neoclassical economics.[26] As an aside, though, I believe that the question of definition on this topic is not much more than a semantical quibble. Whether a set of propositions purportedly descriptive of some portion of the empirical world is referred to as a scientific "model," scientific "program," or otherwise is of little import given that the key term in this context is "scientific." Any scientific model or program must necessarily conform to criteria of testability, prediction, and explanation. I doubt that Weintraub would object to the notion that the general equilibrium model is a purportedly scientific model.

Again, according to Weintraub, only derived theories on the belt of this research program are subject to test and corroboration. The Arrow-Debreu-McKenzie hard core of this program is not subject to critical analysis in terms of its falsifiability potential and corroboration because (1) it constitutes the program's hard core, and (2) it is still in the process of maturation, hence not fully subject to rigorous analysis.[27]

I believe, however, that Weintraub's adaptation of Lakatos's methodology as an explanation of general equilibrium is subject to criticism for the following reasons. Although Lakatos did claim that a research program's hard core ought to be insulated against criticism, he argued nevertheless that a scientific research program is progressive or degenerative according to its surplus or lack of empirical content vis-à-vis its rivals. According to Lakatos, the scientific evaluation of a research program is determined by the experimental tenor of its protective belt. But recall that the protective belt of a research program contains empirically testable theories that entail the hard-core propositions of that program. This means that no empirical test of the protective belt of a theory in a given research program is possible without testing at the same time that theory's hard core.

Thus, while a theory's predictive success reinforces its hard core, its experimental failure questions the same. In this regard, therefore, Weintraub's attempt to defend the general equilibrium theory from criticism fails because that theory's hard core is indeed subject to examination and criticism because of the predictive and explanatory failures not only of the research programs belt theories but also of the general Arrow-Debreu-McKenzie model.

It follows, therefore, that the hard-core propositions of Weintraub's general equilibrium model—that (1) agents independently optimize subject to constraints, and (2) agents have full relevant knowledge—are indeed subject to analysis and criticism. And it is the case that these propositions are empirically dubious. Again, one must question Weintraub's negative heuristic directive: "Do not construct theories in which irrational behavior plays any role." First, the term "irrational behavior" has not been defined, and there is no consideration of whether agents do engage in such. Clearly, if irrational behavior is empirically certifiable, then it cannot be excluded from consideration in a putatively scientific theory.

On examination of Weintraub's formulation of the general equilibrium theory's hard core, it is evident that the term "rational behavior" has not been explicitly defined there. I suspect that it is implicitly embedded in Weintraub's hard-core proposition that "agents independently optimize subject to constraints." This is a universal proposition founded on some principle of optimization that, in turn, derives from the neoclassical postulate of rationality. The reason is that (1) Weintraub speaks explicitly of "irrational behavior" in one of his negative heuristics, and (2) it is an empirical fact that not all agents optimize according to the neoclassical theorist's notion of rational choice.

Furthermore, Weintraub's prescription that the hard-core propositions of general equilibrium theory may be questioned only by outsiders and should command unquestioning assent from its adherents, strikes one as being more appropriate for a metaphysical theory rather than for a scientific discipline. An important methodological rule of scientific investigation is that if a theory fails in terms of its predictive and explanatory claims, then it should be critically examined in its totality (hard core, protective belt, heuristics, etc.) not only by rival theorists but also by its adherents. Otherwise, scientific research reduces to dogma and ideology. I admit, though, that an effecting of this rule is not always easy, given "internal-external" issues, human psychology, and the context boundedness of knowledge. But these are constraints to be borne in mind by the methodologist of scientific research efforts.

In sum, the positive appraisals of neoclassical general equilibrium theory on the part of Coddington, Hausman, Weintraub, and others are subject to the same criticisms leveled against neoclassical microeconomic theory in the sense that the basic axiomatic assumptions posited for general equilibrium theory derive from the normative postulate of rationality. The general equilibrium theory is a highly formalized prescriptive mathematical structure, not a scientific theory with identifiable cognitive content.

But in general, the neoclassical theorist does not recognize this for reasons best understood from the standpoint of the sociology of knowledge. He or she is intellectually conditioned to reify a concept that would appear patently noncognitive to the epistemologist. Thus, in spite of criticism of the lack of explanatory and predictive power of the general equilibrium model, most neoclassical theorists appear to be at a loss to explain why this might be so. This, I believe, derives from an intellectual timidity due to a recognition of the important implications that would result from the jettisoning of the idea that the neoclassical theory is an empirically grounded research program. And to add that the neoclassical theory in general and the general equilibrium theory in particular are founded on normative structures would pose intellectual problems to the large number of theorists nurtured on the long tradition of neoclassicism.[28] The point is that entrenched paradigms seek to prolong themselves, not hasten their own demise.[29] I shall explore these points in the following chapter.

7
The Postulate of Rationality and Neoclassical Economic Theory

Rationality

The concept of rationality is so useful in the interpretation of human experiences that it is invoked not only in matters of epistemological interest such as the nature of scientific inquiry or the status of ethical choices but also in matters of practical interest. Thus, an individual who acts in ways viewed as odd would ordinarily be considered irrational. One also hears the term "irrational" often used to describe the behavior of individuals deemed clinically insane or psychologically disturbed. It is this lower-bound limit of rationality that would appear to imbue it with cognitive content. Yet it is true that in the area of human decision making the subjective choices of means and ends introduce an evident normative content into individuals' schedules of choice. For although there may be justifiable epistemological debates concerning the means chosen to attain certain ends, it should be apparent that there could be no resolvable debate about ends. It would seem that for every argument concerning ends equally valid counterarguments could be generated.

In response to those who would argue that there is a possible cognitive content to the idea of rational behavior because clinically unsound behavior is viewed as being synonymous with irrational behavior, I argue that rationality necessarily entails prescriptivism since its components of means and ends are required assumptions for coherent thought. It would appear, on the contrary, that the clinically unsound are not properly able to identify means and ends in their deliberations. I admit, of course, that there are borderline cases that hinge on definitional issues well-known to legal practitioners. I shall not pursue such questions here since my concern is with the notion of rationality as understood by the neoclassical economist. I reiterate, though, that as long as there is evidence of coherent thought, and means and ends are identified in any decision-making process, then the appeal to the idea of rationality in this connection is to appeal necessarily to some form of prescriptivism.

Although the idea of rationality may be definitionally problematic for the majority of the social sciences, this is not the case with neoclassical economics. Rationality in the economic context reduces simply to the maximization of utility or profits in the most efficient way subject to the agents' monetary and other constraints. This is clearly a case of means-ends or instrumental rationality. But this definition of rationality, it seems to me, is contextual and functional only for a particular kind of economic system.[1] This would mean that a proper explanation of the neoclassical agent's choice patterns would require the metaanalysis of the sociologist or anthropologist. The debating point here is whether genuine evidence exists that human beings are intrinsically utility or profit-maximizing beings. That question has been answered, however, by the positing of certain basic assumptions about human behavior on the part of the theorist of neoclassical economics. These assumptions are usually embodied in the concept of a reified postulate of rational choice that has its roots in a general positivist epistemology.

Consider philosopher of science Carl Hempel's thesis that the concept of rationality may be invoked to explain agent choice. Although Hempel recognizes that the concept of rationality may be used as a normative critical concept, he argues nevertheless *contra* fellow philosopher William Dray that it can explain human choice in the scientific sense.[2] In response to the question "why did A do x," Hempel argues along the following lines:

(1) A was in a situation of type C.
(2) A was a rational agent.
(3) In a situation of type C, any rational ... therefore, A did x.[3]

The above construal differs, according to Hempel, from Dray's in that the premise "A was a rational agent" has been added. He writes further that "in order to ensure the explanatory efficacy of a rational explanation we found it necessary to replace D's normative principle of action by a statement that has the character of a general law. But this restores the covering-law form to the explanatory account" (p. 471). It appears, though, that Hempel's construal hardly effects the qualitative transformation he seeks, since premise (2) contains the epistemologically problematic term "rational." Replace "rational" with the ethical term "bad" and the effect is the same. Hempel's schema could work only if "rationality" were descriptive of some innate human disposition to effect certain kinds of choices and to eschew others. One thinks in this regard of the instinctual drives or dispositions of nonhuman animals with respect to certain aspects of their behavior.

Were Hempel's argument sound, then the cognitive question concerning ethics and other areas of value inquiry would be qualitatively similar to that concerning empirical science. Hempel's epistemological *faux pas* derives from the persistent tendency to reify the concept of rationality. The same holds for neoclassical economics theory for which the concept of rational behavior is expressed as the postulate of rationality. The following section will explore rational choice as defined by the postulate of rationality.

Neoclassical Economic Theory and the Postulate of Rationality

Any study of contemporary neoclassical economic theory would reveal the central role the postulate of rationality plays in that theory.[4] In the previous chapter it was seen how this postulate defines the rational economic agent as one who conforms to certain basic axioms (these, recall, guarantee consistency of choice, granted complete knowledge of the marketplace, and noncyclic preferences) and is an infallible maximizer of utility and profits. But I should want to pursue further in the course of the ensuing discussions the notion that the term "rationality" is conceptually cognate to other noncognitive terms such as "goodness" and "rightness."

Given the axiomatic and mathematical sophistication of neoclassical theory and the scientific claims of its theories, one would expect that its derived models would be empirically testable in terms of their capacity for explanation and prediction. But it is commonly recognized that the record of neoclassical theory as a scientific enterprise has been less than encouraging.[5]

I will argue now that the reasons why neoclassical economic theory in its actual formalization cannot be regarded as scientific are as follows. (1) The postulate of rationality logically insulates its derived axioms, laws, and theories from empirical refutability. On account of this, the theories of neoclassical economics are to be regarded as deductive concatenations of normative ideal-type propositions. (2) The explicit or implicit *ceteris paribus* provisos that serve as auxiliary hypotheses to neoclassical economic models effectively insulate these models from possible falsification. These models can, therefore, be regarded as normative models. (3) The empirical findings relevant to economic theory that could lead to theory reformulation are generally ignored by neoclassical economics theorists, given their *a priori* predilection for purely formal work. The reason for this is that none of the axiomatic propositions and derived theorems is translatable into a purely empirical statement, given their dependence on the normative postulate of rationality and the *ceteris paribus* proviso.

Consider the following theorem. If commodities Q^1 and Q^2 are substitutes, and the consumer's purchases are restricted by a given budget constraint (i.e., he remains on his highest indifference curve), then an increase in the price of Q^1 will induce the consumer to buy more of Q^2 and less of Q^1. This yields:

$$\frac{\partial q_2}{\partial p_1} > 0 \text{ for substitutes}$$

$$U = \text{constant}$$

and

$$\frac{\partial q_2}{\partial p_1} < 0 \text{ for complements}$$

$$U = \text{constant}$$

But given the initial and original posit of neoclassical microeconomic theory that the choices of the model's agents are determined by the postulate of rationality, then the theorems and laws that describe agent behavior are *definitional*, not empirical. In other words, the general propositions about consumers and maximizers in neoclassical theory may be paraphrased as follows:

(1) If A is a rational agent, then A's choices will be transitive.
(2) If A is a rational consumer, then A maximizes utility subject to relevant constraints.
(3) If A is a rational entrepreneur, then A maximizes profits subject to relevant constraints.
(4) If A is a rational agent, then A's choices in economic space conform to Slutsky's equation..., etc.

Empirical cases that do not conform to the predictions of the model are discountable since they do not fall within its context. The neoclassical model is concerned only with behavior that is formulated as rational.

The kind of implication assumed in the above conditional statements is not a material implication but a definitional implication. Evidently rationality is both a necessary and sufficient condition for utility and profit maximization. Thus, to take one of the above examples: If A is a rational agent, then A's choices are transitive implies, in this regard, its converse. But the model also assumes that the antecedents of the conditional statements describing the attributes of rational agents are true of the model. In other words, the model *posits* the existence of rational agents. To put it more formally, one can say that the universe of discourse of the neoclassical model is populated only with rational agents.

Alexander Rosenberg argues, however, that if theorists of microeconomics assume that individual agents are necessarily rational, then most of the propositions of neoclassical microeconomics theory would be analytic. Consider the following passage:

> If economic agents are necessarily rational, that is, if entrepreneurs necessarily maximize profits and consumers do the same for utilities, then most of the general propositions about them not only follow deductively from these statements, but are analytically true as well. Why should this be so? Microeconomic statements are fully justificatory; that is, given the choice under the circumstances assumed by (classical) microeconomic theory, their antecedents justify one and one action as rational. They do so because it is an analytic truth that a fully rational agent will perform the action in question under the circumstances.[6]

Yet Rosenberg claims that "economists do not assert that agents are rational by definition, nor is it legitimate to infer analyticity from unfalsifiability."[7] The argument that Rosenberg invokes in his attempt to demonstrate that the behavioral assumptions of the neoclassical model are those descriptive of rational agents is that there is empirical evidence that not all agents conform to the postulate of rationality, and that when this postulate is violated agents will tend to change their behavior to conform to it.[8] But this is precisely the point: the neoclassical theory remains entrenched despite falsifying empirical evidence. Rosenberg argues, however, that rationality is a property of individuals according to the theoretical assumption of neoclassical theory. Thus, again, the neoclassical economic model is not founded on the observed behavior of actual agents, but on the derived choice paths of the rational economic agent. I should want to argue, however, that Rosenberg's claim that rationality is a property of individuals is erroneous. I also take issue with the claim that those propositions founded on the postulate of rationality are analytic. My counterclaim, rather, is that those propositions which constitute the theoretical structure of the neoclassical model are evaluative in structure.

The proposition that "x is a human agent" is a proposition that could be empirically confirmed, since questions concerning the "humanity" of the agent in question could be answered by appeal to biological evidence, even if such evidence were established by definition according to a set of necessary and sufficient criteria. But "rationality" is not one of such criteria. It is indeed possible that there be irrational human agents. Rationality, as defined by the neoclassical postulate of rationality, requires that the agent conform to a set of axioms that would guarantee a set of optimal choices, whether maximal or minimal—granted certain real constraints. Surely, such a model descriptive of ideal behavior should be best viewed as evaluative. One could just as easily construct an ethical theory in which the term "good" is substituted for "rational." There will indeed be confirming instances of "good" behavior, according to the theory, but such confirming instances do not resolve the problematic status of "good." For purposes of empirical analysis, terms such as "good" and "rational" are provably redundant for investigatory purposes.

Similar considerations apply to the claim that the neoclassical theory is analytical in structure. The claim that some proposition P is analytical in structure is determined by its logical status, specifically whether it may be denied without contradiction. Thus, the proposition "$6 + 4 = 3 + 7$" cannot be denied without incurring a contradiction. The logical status of an analytic proposition is also determinable purely by the meaning of its constituent terms. Thus, the proposition "all three-sided two-dimensional figures are triangles" is an example of an analytic proposition purely in terms of the meaning of the terms "three-sided two-dimensional figure" and "triangle." But recall the logician Quine's skepticism concerning the claim that there is some easily identifiable distinction between analytic and synthetic propositions.[9] In the same vein, some theorists have expressed uncertainty as to whether some propositions usually viewed as synthetic or empirical are to be so regarded. Newton's second law, $F = MA$, is viewed by some as analytic despite its exclusive usage as an empirical law. I am inclined to believe that Quine's analysis does have merit, but I would also argue that propositions generally regarded as empirical laws cannot be viewed as analytic since the basis for their formulation is empirical evidence. In the case of Newton's second law, the idea of "MA" is not immediately and necessarily derivable from the notion of "F." Furthermore, it is not at all impossible that at a future date some researcher could discover that there are possible empirical conditions for which $F = MA$ is not a true statement. I do not perceive the same eventuality for purely definitional statements that establish *a priori* semantical equivalences between terms.

However, when the claim is made that the structure of neoclassical theory is analytical, it would appear that what is meant is that its constituent propositions are logically true given their mathematical structure. And since the time of Kant, few theorists have argued that formal and provable mathematical propositions are not merely analytically true and incapable of being shown false. But it is certainly difficult to claim that "rational agent" is logically equivalent to "maximizer of utility or profits," according to the orthodox test for analyticity. The point is that logical equivalences between terms are provable only on condition that the meanings of the terms in questions be clearly stated and understood. It has been shown, though, that the meanings of the terms "rationality," "rational behavior," and so on are not yet clearly defined. Consider decision scientist Patrick Suppes's observations:

> Recent work in decision theory has shown that there is no simple coherent statement which corresponds to naive ideas of rationality. Just as research in this century in the foundations of mathematics has shown that we do not yet know exactly what mathematics is, so the work in decision theory shows that we do not yet understand what we mean by rationality. Even in highly restricted circumstances it turns out to be extremely difficult to characterize in a nonparadoxical fashion a rational choice among alternative courses of action.[10]

Or witness Amartya Sen's statement that "if today you were to poll economists of different schools, you would almost certainly find the coexistence of beliefs (i) that the rational behavior theory is unfalsifiable, (ii) that it is falsifiable and so far unfalsified, and (iii) that it is falsifiable and indeed patently false."[11] Clearly, the unanswered questions concerning the postulate of rationality, so central to neoclassical theory, is further evidence of its problematic status as a scientific proposition.

Given the above, I argue that the postulate rationality is best viewed as a normative proposition than otherwise. But I should want to show, too, that a jettisoning of this postulate would create a situation whereby the theorist would be incapable of formulating any meaningful theory of economic behavior. The implication of this is that the theorist is obliged to accept the fact that economics as a research discipline is necessarily value laden.

But consider Machlup's attempt to resolve the problematic status of the postulate of rationality by claiming that its derived theoretical assumptions and postulates should not be evaluated purely in terms of their empirical content but should be viewed rather as heuristic principles or procedural rules necessary for the interpretation of economic data. The basis for this interpretation of the rationality postulate is that the theory derived from it works well in a wide variety of circumstances, despite evident anomalous cases.[12] In this regard, the postulate of rationality as embodied in the ideal type of a figurative "economic man" should be viewed as a necessary assumption for economic theory.[13] Machlup's rationale for the theoretical role of this construct is not to

> watch anyone's movements and gestures nor listen to their conversations—but rather to understand observations of records, such as reports on prices, outputs, employment, and profits which are evidently *results* of people's actions and reactions. Almost never can we observe the actions themselves.[14]

It is on the basis of this ascription of a heuristic function to the rationality postulate that Machlup can claim that the fundamental assumptions[15] of neoclassical theory "though empirically meaningful require no independent empirical tests but may be significant steps in arguments reaching conclusions that are empirically testable."[16] Machlup also claims that the fundamental assumptions of physical mechanics are likewise impervious to independent verification.[17] I harbor doubts about this claim on the grounds that there would be no epistemological quarrel with Machlup's instrumentalist approach were the predictions of neoclassical theory borne out—as they are with a properly constrained physical mechanics. On the other hand, there is sufficient anomalous evidence to question seriously the adequacy of the postulate of rationality as a basic untestable assumption of neoclassical economic theory.

Machlup himself argues that the heuristic approach to the postulate of rationality should be maintained until displaced by a rival theory not using that postulate. Such a theory could indeed be forthcoming were Machlup to view the postulate of rationality as essentially prescriptive. But Machlup rejects this view in his elaboration of the idea that there is an evident epistemological distinction between positive and normative economics.[18] Thus although, as he put it, the "positive" in positive economics is not the equivalent of "observable," it nevertheless means nonnormative and nonevaluative.[19]

Methodologist of economics Bruce Caldwell would prefer, however, to support Machlup's thesis that the rationality postulate (as a fundamental assumption of neoclassical economic theory) is not subject to direct test for its validity status.[20] Caldwell's thesis is that since the initial conditions of the neoclassical theory are subject to change because of the changes in tastes and preferences on the part of the individual agent, any test of the rationality assumption would not yield confirming or disconfirming results. But this is an odd claim to make since the rationality postulate has indeed been tested for its empirical reliability with results that have not been sufficiently encouraging. One might consider in this regard the well-known research efforts of Kahneman and Tversky[21]: individual agents do not always maximize expected utility as the postulate of rationality predicts.

Granted the importance that researchers in science ascribe to empirical analysis, one must take issue, too, with Caldwell's endorsement of Machlup's thesis that "empirical studies can never establish nor falsify a theory, but can be used to judge its applicability."[22] For if this were the case, the question immediately arises of how rival scientific theories could be evaluated. The standard and well-proven scientific practice has been to choose between theories in terms of their respective explanatory and predictive capacities according to the available empirical evidence. In fact, this idea is implicitly borne out by Caldwell's endorsement of Machlup's claim that a disconfirmed theory may not be replaced until another satisfactory theory is formulated. The point is that a scientific theory may be disconfirmed (assuming that it is logically sound) only on empirical grounds. Clearly, the Machlup-Caldwell claim that "empirical studies can never establish nor falsify a theory" is not supportable.

But let us pursue the thesis that neoclassical economic theory is avowedly scientific in intent and that its rationality postulate portrays the choice paths of existent agents and argue that a possible solution would be to formulate a theory of decision making that could incorporate all kinds of choice schedules. And each of these schedules would be founded on its own particular postulate of rationality. The immediate problem, though, would be that the great number of possible choice schedules would make theory construction unmanageable. As logician Frederick Schick put it, "whether a person is choosing rationally depends on his understanding of things—this I will label the relativity (of rationality) to understandings."[23] Schick's interpretation of rationality, as it applies to expected utility theory, is that utilities should be viewed intentionally, that is, as they bear on individual understanding. The point is that "different people may see things differently, and the utilities that figure on each reflect that person's own understandings."[24] Yet, there is a practical solution to the epistemological impasse confronted by the theorist because of the multiplicity of possible decision-making schedules. An appropriate solution, it seems to me, is to apply Occam's razor and to construct a theory of optimal decision making and to recognize it as a purely prescriptive argument.

Yet, another solution could be to dispense altogether with the notion of rationality, but then theory construction, according to orthodox neoclassical economics, would not be possible. Recall that the failure of the cardinal utility approach to economic decision making forced theorists of economics to posit the postulate of rationality as an *a priori* assumption on which explanatory theories could be constructed. Otherwise, the theorists, unable to measure mental states as the explanatory factor of human decision making would be constrained to a mere taxonomic data gathering. The theorist could, of course, observe behavior from a strictly anthropological viewpoint and record not only the choices made by individuals but also the model of decision making to which they appealed in order to effect those choices. The theorist, after taking an adequate sample, could arrive at the conclusion that a given model of decision making was prevalent (i.e., culturally embedded) within society.

But from a strictly scientific point of view, this would not meet the requirements for a genuine scientific theory. A genuine scientific theory is much more than the descriptive formulation of a set of empirical facts; it must go beyond the facts and offer explanations and predictions. In order to achieve this, however, the theorist must appeal to some model of decision making that could support counterfactual conditionals. But a model that supports counterfactual conditionals must appeal to deterministic principles. Given the problematic nature of human behavior, one solution the theorist could adopt would be to construct a theory founded on the observed prevailing decision-making theory. This theory would, of course, defeat its own purpose. It would not be a scientific theory—it would be best construed as evaluative.

Thus, in sum, I do not believe that it is possible to construct general theories of economic decision making without appealing to some *a priori* model of rational choice. The alternative would be to formulate a model of decision making founded on a theory of mental states. And I have shown that this approach, as demonstrated by the cardinal utility thesis, is scientifically indefensible. But this discussion has also shown that the alternative approach by appeal to the postulate of rationality is also scientifically unacceptable since the axiomatic structure supported by this postulate is evaluative in content.

It is for this reason that the claim by economist Clem Tisdell that the rationality assumption "can be used to predict a good deal of human behavior, on average and within certain limits"[25] cannot be viewed as proof of the scientific tenor of neoclassical theory on the basis of the above discussion. The "good deal of human behavior" to which Tisdell refers would, no doubt, be restricted to behavior in those societies where the neoclassical model is dominant. It would be difficult to accept the thesis that the orthodox neoclassical postulate of rationality could adequately explain and predict behavior in devout Buddhist societies. Tisdell has obviously failed to recognize that the neoclassical postulate of rationality specifies not only "means" but also "ends." And these "means" and "ends" are not freely chosen by the agent himself but formulated for him in catechistic fashion by the neoclassical theorist. Just as the law and society evaluate the acts of individual agents in terms of their ethical content—a content formalized in textbooks of sociology, psychology, and psychoanalysis by the usage of terms such as "deviant" and "nondeviant" behavior—so, too, the economic decisions of individual agents, though evaluated by the public in terms of "success" or "lack of success," are viewed by the neoclassical theorists of economics as instances of "rational" or "irrational" choice.

But that is just the nature of human behavior: human behavior is potentially free and nondeterministic, hence potentially dysfunctional for organized social collectivities. This explains the universal human inventions of prescriptive codes for ethical, legal, and economic behavior. The result of this is that the theorist cannot avoid formulating theories that are necessarily normative. Frank Hahn and Martin Hollis, for example, argue that if the pure economist seeks rational solutions for economic problems, then he must construct explanatory theories founded on some concept of rationality. Consider the following observations:

> The term "rational" occurs essentially in *a priori* statements whose truth is crucial to the descriptive, explanatory, and prescriptive uses of the theory. So it does matter that the theory is concerned with the concept of rationality and whether it is right about it. Otherwise, the pure economist is rational neither in his choice of assumptions nor in his advice.[26]

Yet, most theorists of neoclassical economics would not consider, in the light of the present analysis, giving up the neoclassical paradigm. Nor would they be willing to concede that the distinction between positive (scientific) economics and normative (welfare) economics cannot be justified since the role of the postulate of rationality in orthodox neoclassical theory confers on it an unavoidable normative status. Claims persistently will be made about the testability of portions of neoclassical theory. There will be little recognition of the fact that the choices of "rational" economic men are no more scientifically testable than the acts of, say, "good" men. I will continue to challenge this oversight in the ongoing discussion.

The evident theoretical problems generated by the classical deterministic model of rational economic man as an omniscient calculator has prompted some theorists to propose alternative theories of rational decision making more compatible with the actual processes of individual decision making. Among the most noteworthy of these efforts are the theories of bounded and procedural rationality developed by Herbert Simon.[27] Simon's thesis is founded on the idea that the concept of rationality employed by neoclassical theory does not take into account the many *faux pas* and uncertainties of actual decision making. This theory of rationality is described by Simon as substantive in the sense that "it is appropriate to the achievement of given goals within the limits by given conditions and constraints."[28] Simon argues that a plausible solution to the problem of rational decision making in economic theory is to borrow the psychologist's idea of procedural rationality. According to Simon, "behavior is procedurally rational when it is the outcome of appropriate deliberation. Its procedural rationality depends on the process that generated it."[29] Simon explicates this concept further by claiming that procedural rationality is not concerned with the formulating of optimal solutions, but rather with the establishing of step-by-step computational procedures to find the best practical solutions.[30]

Simon concludes his thesis by claiming that the deductive axiomatic model of substantive rationality is less suited to the interpretation of human decision making than the empirical analysis of complex algorithms of thought by appeal to the concept of procedural rationality.[31] The point is that any theory of rational decision making must take into consideration the real constraints and cognitive limitations of the agent. In fact, it appears that any adequate theory of rational decision making must appeal to background learning and feedback theories.

C. A. Tisdell takes issue with Simon's criticism of the neoclassical model of unbounded and substantive rational decision making by arguing that the strength of this model derives from its affording the best strategy for the agent to choose.[32] But to refer to a strategy as optimal is to admit implicitly that it is indeed directive. Similar considerations apply to Simon's attempt to substitute the idea of procedural rationality for substantive rationality.

But consider, too, the attempt by economist Gary Becker to defend the role of the postulate of rationality on the grounds that this postulate describes better the behavior of the market as a whole than the behavior of individual agents. On this point, Becker writes that:

> Today, critics either deny that households maximize any function or that the function maximized is consistent and transitive. In effect, they deny that households act "rationally" since rational behavior is now taken to signify maximization of a consistent and transitive function.[33]

Becker argues that markets may be rational in the sense of, say, having negatively sloped demand curves, while individual decision makers may not be rational. Thus, according to the same author, "the important theorems of modern economics result from a general principle which not only includes rational behavior and survivor arguments as special cases, but also much irrational behavior" (Becker, p. 153). Yet, he also claims that by appealing to the idea of changing "opportunity sets," it can be demonstrated that "irrational" decision makers could be made to revise their choices and, henceforth, choose "rationally" (p. 167).

However, there are three problems with this approach to vindicate the role of the postulate of rationality in neoclassical theory: (1) An emphasis on market behavior as a possible heuristic solution to the problem of the postulate of rationality would tend to offer no understanding of the decision making mechanics of the individuals in the economy: firm and individual agent; (2) a holistic approach to economic behavior would tend to promote the formulation of descriptivist or taxonomic models of analysis. Science, however, is concerned not only with description but also with explanation. The explanation of the behavior of the market must be sought essentially in analyses of the decision making choices of its individual units; and (3) the claim that irrational units can be made to make rational choices by a change in opportunities is not borne out by the facts: how can the theorist explain the common phenomenon of bankruptcy except to argue that the decision maker has persisted in making "irrational" choices.

I conclude this section by stating that the formulation of any predictive and explanatory model of decision making for agents who make choices based on conscious deliberation would be based on normative considerations. Human choice making, by definition, requires the subjective positing of goals to be attained and the means of so doing. Any theory descriptive of such would be normative while the actual choices made by agents, though empirical, would be explainable only in terms of the normative model in question. The useful aspect of Simon's thesis is that it points out that actual agent behavior could diverge radically from its relevant theoretical model. The neoclassical theory and its concept of substantive rationality do not make that distinction.

Yet, the important point to note is that whatever model of rational choice is adopted for economic decision making, the implications are the same: such a model must be founded on normative principles. The evident biological given is that human agents are not guided by instincts in their decision making but by conscious deliberation. This conscious deliberation can be guided only by normative programs of means and ends. A strict empirical analysis of behavior without appeal to some normative explanatory model would be incomplete in the sense that a full explanation of human decision making must appeal to its causes, that is, motives. But motives are not instinctual phenomena; they derive from specific forms of cultural conditioning, though they appear to be freely chosen. Thus, whatever model of explanation the theorist formulates to analyze choices would be derived from normative assumptions given the great variety of other models that any agent could consciously adopt.

The *Ceteris Paribus* Proviso

Predictive laws or theories in neoclassical theory are often qualified by a *ceteris paribus* proviso. This proviso is usually employed in the following manner: "If the price of a commodity increases, then *ceteris paribus*, the consumer will demand less of its complement." If one categorizes the antecedent of the above proposition A and the consequent B, then it may be categorized as follows: "If A, then B, all other conditions being equal."

Questions have been raised about the specific role played by the *ceteris paribus* proviso in neoclassical theory. Hutchison, for example, argues that the use of the *ceteris paribus* proviso in neoclassical theory effectively "makes out of an empirical proposition that is concerned with facts, and therefore conceivably can be false a necessary analytically-tautological proposition."[34] Mark

Blaug argues, however, that the problem concerning the status of the conditional type statements of economic theory does not derive from usage of the *ceteris paribus* proviso since it is also employed in the physical sciences. Note the following:

> A scientific theory that could entirely dispense with *ceteris paribus* clauses would, in effect, achieve perfect closure: no variable that makes an important difference to the phenomena in question is omitted from the theory and the variables of the theory, in effect, interact only with each other and not with outside variables.[35]

But Blaug fails to note that in the case of the physical sciences, the conditions that might constitute *ceteris paribus* assumptions are usually stated explicitly. In fact, it is for this reason that, contrary to Blaug's claims, the *ceteris paribus* proviso is hardly ever employed in theory formulation in the natural sciences. Hempel makes the same point when he writes:

> It is significant to note here by contrast that in the formulation of physical hypotheses, the ceteris paribus clause is never used: all the factors considered relevant are explicitly stated (as in Newton's law of gravitation or in Maxwell's laws) or are clearly understood (as in the familiar formulation of Galileo's law, which is understood to refer to free fall in a vacuum near the surface of the earth); all other factors are asserted by implication, to be irrelevant.[36]

Yet, although it is evident that the *ceteris paribus* proviso in neoclassical theory does refer specifically to the assumptions of the theory, the reason why it is not replaced by explicit statements as in the case of the natural sciences is that while in the natural sciences the *ceteris paribus* clause refers to empirically certifiable and manipulable propositions, in neoclassical theory the proviso refers not to empirical data but to the uninterpreted axiomatic assumptions of the theory. These axiomatic assumptions are normative propositions derived from the postulate of rational choice. Thus, in the context of neoclassical theory "if A then *ceteris paribus* B" would translate as "if X is a rational neoclassical agent, and all of the theory's background assumptions hold, then if A then B." The *ceteris paribus* proviso, it seems, is merely about the normative content of the axiomatic structure of neoclassical theory.

Blaug argues that the unspecified nature of the *cetera* in *ceteris paribus* renders *ceteris paribus* clauses nontestable, even though such clauses are purportedly empirical.[37] It was argued above, though, that the *cetera* in *ceteris paribus* is indeed specifiable; it refers to axiomatic assumptions of the theory and contains only statements derived from the normative postulate of rationality. But even if one examines Blaug's notion that *ceteris paribus* clauses are nontestable even if empirical, then, the reason for such is clear. The complexity of human action and the unpredictability of empirical circumstances can never be adequately formulated by the theorist unless he or she possessed mental and intellectual powers far beyond those of the standard neoclassical theorist. Even within the context of highly constrained theories such as those founded on the concept of bounded rationality, the possible instantiation of the *"cetera"* in *ceteris paribus* would still remain elusive.

Normal Science and Neoclassical Economic Theory

Although economic theory deals with empirical facts, much of its data is ignored on the grounds that it does not fit the model in question. Consider Blaug's observations:

> instead of attempting to refute testable predictions, modern economists all too frequently are satisfied to demonstrate that the real world conforms to their predictions, thus replacing falsification, which is difficult, with verification, which is easy.[38]

The matter is compounded by the fact that the economics profession assigns most status and prestige to research that refers only incidentally to the empirical world.[39]

In this regard, normal scientific work in neoclassical economic theory tends to follow the methodological prescriptions of the formalist paradigm. Thus, it is not of much concern to orthodox neoclassical theorists that the correspondence between theory and empirical fact is minimal. Yet, this is not to argue that neoclassical theorists deliberately choose not to test their theories against the empirical world. One encounters here a situation in which a group of researchers formulate theories from within the confines of a given research paradigm.

In the case of research paradigms in the empirical sciences, the path and speed of progress and the generation of competitiveness between research programs is determined by the explanatory and predictive power of the programs in question. But the research programs in neoclassical economic theory do not stress empirical prediction, explanation, and falsification as determinants in theory evaluation. Rather, formal rigor is what is viewed as a necessary and sufficient condition for theory acceptability. The empirical content of a theory is of minor concern for the orthodox neoclassical theorist.[40]

Yet, unfortunately, those theorists who, in instrumentalist fashion, determine the tenor of their theories by their predictive results fare no better since the empirical predictions of neoclassical theory are not, in general, borne out in fact.[41] The reason for this is logically evident: the axiomatic foundations of neoclassical economic theory represent an ideal-type framework that offers normative prescriptions for economic decision making.

On account of the weak predictive record of neoclassical economic theory, some theorists have relaxed the assumption of a perfectly transparent market in order to accommodate the possibility of the economic agent making choices in uncertain situations. For example, in an appraisal of formal deterministic agent-choice theory, Henderson and Quandt write: "The previous analysis is unrealistic in the sense that it assumes that particular actions on the part of the consumer are followed by particular determinate consequences which are knowable in advance."[42] The solution proposed by the authors is "to construct a utility index which can be used to predict choice in uncertain situations if the consumer conforms to the following five axioms."[43] The reference here, of course, is to the expected utility theory, which has been discussed in a previous chapter. I reiterate here that the expected utility approach is problematic since its axiomatic foundations are conceptually similar to the normative axioms of deterministic neoclassical theory.

A Postscript on Rationality

In the above discussion, I have attempted to analyze the concept of rationality in a fairly systematic way. I say this because most texts on the scientific status of economic theory do not focus much on this concept. Those theorists who choose to discuss it tend to be rather apologetic for its usage. I refer here to authors such as Machlup and Caldwell.[44] However, there have been two noteworthy attempts to examine the concept of rationality as the central directive concept of neoclassical economic theory. In this section, I believe it instructive to comment on these two research efforts. They are Hollis and Nell's *Rational Economic Man*[45] and Amitai Etzioni's recent *The Moral Dimension—Toward a New Economics*.[46]

Hollis and Nell engage in a concerted attack on neoclassical economic theory and its central concept of rational economic man, which they view as an artificial concept founded on logical positivism. But their solution, it seems to me, is even more problematic than the claims of neoclassical theory itself. They propose a rather nebulous rationalistic approach to economic theory founded on the theories of classical and Marxian economics. This rationalistic approach derives from the notion that maximizing behavior (hence, some form of the postulate of rationality) is central to economic theory.

More specifically, on the issue of rational economic man, the same authors, though critical of the neoclassical version of such, propose nevertheless the curious claim that "there are rational economic men" is a synthetic *a priori* proposition.[47] They seek, furthermore, to retain the nebulous concept of rationality by arguing for what they regard as a weaker version. "Rational action," according to Hollis and Nell, is defined as "regularity of behavior." They elaborate on this by arguing that the principle of maximization provides the moving force of economics.[48] The problem is that though the authors engage in a forceful critique of the neoclassical concept of rational choice, they nevertheless retain this concept for the formulation of their own classical-Marxian approach. This, no doubt, is testimony of the deep entrenchment of this concept.

Of much interest, too, I would think, is the research effort of Etzioni, who points out that the neoclassical theory of rational behavior is overly restrictive in its description of actual human economic behavior. Etzioni argues that rationality ought not to be restricted only to the calculating behavior of individual utility maximizers, but should be understood as also being grounded deontologically in the social ethic. In short, Etzioni's claim is that individuals make their decisions not only with considerations of efficiency in mind, but also with regard to the ethical values of society. I believe that Etzioni's analysis possesses some sound points in its critique of the neoclassical theory of rational choice, but I argue that this analysis is deficient in that it does not offer a structural alternative or solution to the problematic of economic decision making.[49] The "black-box" question, I believe, is as follows: if the student of neoclassical economics rejects the neoclassical paradigm, what qualitatively similar approach should be adopted? If normative considerations also determine economic choice, the question then is how might this be expressed in terms theoretically similar to those of the existing neoclassical paradigm?

The point is that although Etzioni offers a useful critique of neoclassical economic theory from the standpoint of the sociology of knowledge, he does not subject it to analysis on terms sufficiently familiar to the neoclassical theorist. As proof of this, consider Etzioni's claim that the neoclassical economics paradigm is bereft of normative factors in its formal structure, though its implications may be normative in content. A proper analysis of the postulate of rationality and its role in economic theory would show that this claim is dubious.

In fact it is just this palpable unwillingness on the part of the neoclassical theorist to embrace the idea of the central importance of the postulate of rationality in the construction of economic theory that affords grounds for the critical analysis of those skeptical of the scientific credentials of this theory. Consider Alexander Rosenberg's[50] argument that "generic predictions" of economic theory are much too vague and must be supplemented with other theories that could confer more predictive (hence explanatory) specificity on this research discipline.

It is this weak structure of the predictions of economic theory that thwarts the attempts by theorists like Philippe Mongin who would seek to vindicate Friedman's "irrealism of assumption thesis,"[51] which, we recall, determines the tenor of a theory crucially by its predictive capacity. No liberal epistemologist of science would fault Mongin's thesis that the "partial interpretation" view of theory construction should entail a controlled freedom in the task of the formulation of models of economic behavior.

But this approach would be acceptable only if the relevant theory were successful in terms of offering specific predictions.[52] It would seem that the persistence in attempting to salvage a theory that has shown itself to be problematic not only in terms of its predictive yield, suggests evident conceptual disorientation on the part of the neoclassical theorist. The formulation of neoclassical theory hinges on the postulate of rationality which properly construed is prescriptive in format.

A fitting conclusion to the analysis in this chapter is that, contrary to its avowed methodology, the neoclassical research program is minimally concerned with empirical analysis and maximally concerned with refinements of the formal models of individual agent theory and general equilibrium analysis. The basic assumptions of neoclassical microeconomic theory are constituents of a normative framework populated only by "rational" agents. This peculiar relationship to the empirical world is determined by the postulate of rationality and the *ceteris paribus* proviso. In the case of a genuine scientific theory, basic axiomatic assumptions are linked deductively to the empirical world by means of bridge principles. It is in this context that one speaks of the predictive test implications of a theory as a confirmatory warrant of its theoretical assumptions. This is not the case when the test implications of a theory are not deductively related to its assumptions—as is the case with neoclassical economic theory.

8
"Positive" Neoclassical Welfare Economics

According to standard methodological analysis, neoclassical economic theory makes a distinction between positive economics and normative economics. Up to this point, I have been concerned to examine the scientific claims of positive economics. Yet a proper evaluation of the scientific claims of neoclassical theory would not be complete without an examination of the role and claims of neoclassical welfare economics within the general context of neoclassical theory.

One of the reasons I have for examining neoclassical welfare economic theory is that I should want to show that the theoretical foundations of orthodox welfare economics are identical with those of positive economic theory. The intentions here are rather interesting, since some formalists of neoclassical theory argue that Paretian welfare economics—which affords the theoretical basis for orthodox welfare economics—should really be viewed as a branch of orthodox positive theory. This argument is indeed valid, but it is not empirically sound because I have already demonstrated that orthodox neoclassical theory is founded on normative assumptions.

The purpose of this chapter, therefore, is to demonstrate that the thesis promoted by some theorists that formal welfare economics should not be regarded as a normative discipline is open to question. It is argued that formal welfare economics, with its emphasis on efficiency and an inability to handle questions of equity, is more closely aligned with orthodox neoclassical theory than is otherwise assumed.[1] The structure of this chapter is as follows. First, I shall show that the claim made by orthodox theorists that welfare economics is founded on normative considerations distinct from positive theory is subject to question. I shall demonstrate that one could not argue logically that positive neoclassical economic theory is bereft of normative content while claiming at the same time that welfare economics is founded on normative assumptions. Given a provable logical connectedness between positive neoclassical economic theory and Paretian welfare economics, the valid claim would be that neoclassical theory as a whole is either normative or positive from an epistemological point of view. But given my thesis that positive neoclassical economic theory is normative in content, my argument will lead, therefore, to the further thesis that formal welfare economic theory is equally normative as positive neoclassical economics theory.

The Question of Welfare Economics

It should be noted first of all that the classical and postclassical economists made little distinction between objective analyses of the economy and prescriptive recommendations about its functioning. This was the basis for the fact that the discipline was known not as economics or economic science, but as political economy. The writings of Smith, Ricardo, and Malthus were naturally concerned not only with the empirical details of economic decision making but also with expressing value judgments about the accumulation and distribution of wealth.[2] However, as economists began to rely more on quantitative methods and to support the belief that an objective, hence, scientific study of economic behavior was possible, especially with the introduction of the cardinal measurement of utility, the idea that evaluative questions were to be treated separately from those concerned with strictly empirical analyses of economic decision making became popular. Contrast, for example, John Stuart Mill's claims that economics is to be viewed as a moral science with the views expressed by Jevons and Walras. The latter were concerned to establish economics as a purely empirical science. And even in those areas where political questions or questions relating to welfare and equity were rightfully raised, the quantitatively minded Pareto saw fit to establish a theory of welfare economics in which efficiency rather than equity was the major methodological concern. Herein lies the basis for the view of Paretian welfare economics as representative of formal neoclassical welfare economics.

Yet the belief is still maintained in orthodox circles that welfare economics represents the normative side of neoclassical theory and that it is fundamentally distinct from positive economic theory.[3] But some authors are puzzled by this distinction, as the following observations by welfare economist Yew-Kwang Ng demonstrates:

> Now, what about welfare economics? While these is no consensus, a majority of economists seem to regard welfare economics as normative.

This seems to be a little curious, as a majority also regard economics as a science. If economics is a science (which is positive), then welfare economics, as a part of economics should also be a positive study. But is welfare economics perhaps not a part of economics? This is an apparent inconsistency.[4]

Stronger claims to this effect were made by Archibald[5] and more recently by Hennipman.[6] In a fairly detailed essay, Hennipman argues that modern welfare economics should be properly regarded as an aspect of positive economic theory. Hennipman also claims that to treat Paretian welfare economics as if it constituted the normative component of economic theory would be to ascribe to it characteristics that it does not have. He writes:

> The presumption, even if qualified to some extent that Paretian welfare economics has the mission to act as society's ethical guide is hollow, and tends to drag it into a morass of deepening confusion and idle polemics. The sooner this pretension is relinquished, the more fruitful the future history of Pareto's offspring will be.[7]

But although most neoclassical theorists would tend to argue that there is a justified basis for the theoretical distinction between positive economics and welfare economics, this distinction is not at all evident given the close similarity between the formal structures of both theories and the operational limitations placed on orthodox (Paretian) welfare economics. I shall explore this claim in the following sections.

The Structure of Neoclassical Welfare Economics

Neoclassical welfare economics is founded on the following basic assumptions, which it shares with positive neoclassical theory. Assume the usual significance of S (set of states of the economy), X (individual states of the economy), and social rankings R (weak), P (strict), and I (indifference). As usual, one derives the following axioms and definitions:

1. Axiom 1: If $X^1, X^2 \varepsilon S$, then either $X^1 R X^2$ or $X^2 R X^1$
2. Axiom 2: If $X^1, X^2, X^3 \varepsilon S$, and if $X^1 R X^2, X^2 R X^3$, then $X^1 R X^3$

3. Axiom 3: If $X^1 \varepsilon S$, then $X^1 R X^1$
4. Definition 1: $X^1 P X^2$ if $X^1 R X^2$ and not $X^2 R X^1$
5. Definition 2: $X^1 I X^2$ if $X^1 R X^2$ and $X^2 R X^1$

Thus, given any social ranking consistent with the properties of completeness, transitivity, and reflexivity, $W = W[R_1(x), \ldots, R_n(X)]$, representative of individual rankings by individual agents, is derivable as a social-welfare function.

Given the above observations and the prohibition against interpersonal comparisons of utility, the key definition about Pareto optimality is derivable given the following assumptions:

1. $X^1, X^2 \varepsilon S$
2. If $X^1 R_i X^2$ for all $i = 1, \ldots, m$, then $X^1 R X^2$
3. If $X^1 R_i X^2$ for all $i = 1, \ldots, m$, and $X^1 P_j X^2$ for some j, then $X^1 P X^2$
4. If $X^1 I_i X^2$ for all $i = 1, \ldots, m$, then $X^1 I X^2$

A social state $X°$ is Pareto-optimal in S if there is no state $X \varepsilon S$ such that $X P X°$. This formulation means that a social state is Pareto-optimal if there is no ranking relationship that could make every individual at least as well off and at least one individual better off. Quite obviously, Pareto optimality is essentially concerned with efficiency rather than equity. The theoretical justification for its structure is the fact that ordinal utility theory does not allow utility measurements and interpersonal comparisons of utility. These prohibitions raise questions as to whether social-welfare functions representing the collective rankings of all individuals in society are feasible. On the basis of the ordinal utility theory, only individual rankings of the social-welfare function are possible.

The point being made is reinforced by Kenneth Arrow's well-known possibility theorem. This theorem states that if interpersonal comparisons of utility are ruled out, then no social-welfare function can be formulated given five conditions guaranteeing individual preferences and nondictational rankings. But going one step further, I argue that if one accepts the idea that interpersonal comparisons of utility are to be ruled out in the neoclassical welfare economic theory, then its prescribed function as that branch of economics which is specifically concerned with questions of equity and the just distribution of wealth is no longer justifiable. But although it loses its normative role in this respect, it nevertheless retains it on the basis of the claim that its axiomatic structure is founded on the normative postulate of rationality and its attendant axioms.

Contemporary neoclassical welfare economics may be viewed, however, as a combination of Paretian welfare economics and the results of Arrow's research on the possibility of establishing a viable and logically derived social-welfare function. One could conclude, therefore, that given the results of Paretian welfare economics and Arrow's theorem, there is much justification for the claim that there is no qualitative difference between the structure of welfare economics and positive neoclassical theory.

Despite the fact that it is evident from this discussion that there is much coincidence between the structures of neoclassical welfare economic theory and positive economic theory, the notion that both components of neoclassical theory are qualitatively distinct is still held because of the original methodological intent of welfare economics. The original purpose of neoclassical welfare economics was to derive proper social-welfare functions founded on normative notions such as general equity and comparable individual welfare. To put it more clearly, the purpose of neoclassical welfare economics was to address questions of social and economic justice.

But Pareto's formal system is not much more than a reinterpretation of the neoclassical ordinal theory—specifically, indifference curve analysis. Recall that indifference curve analysis states that the tangency between the agent's highest indifference curve and his budget line signifies not only the highest level of satisfaction but also the most efficient position. Yet given the proscriptions against interpersonal comparisons of utility, the most that a Pareto ranking could achieve is a formulation in which there are no distributional tradings. For according to Paretian welfare economics, only those rankings for which no agent perceives his welfare to decrease may be formulated. The reason for this is that recommended rankings in which at least one agent is made worse off while another is made better off would entail assumptions of interpersonal comparisons of utility. Yet a genuine welfare economics must grapple with the issue of interpersonal comparisons of utility if it is to deal with questions concerning equity, income distribution, and so on.

The evident limitations of the Paretian theory in answering issues of equity and distribution have led some theorists to raise genuine welfare economic questions concerning appropriate mechanisms of asset transfers and income distribution such as compensation payments and so forth. Consider the Hicks-Kaldor[8] compensation principle, which states that if in two states of the economy Z^1 and Z^2 a move from Z^1 to Z^2 entails gainers in Z^2 possibly compensating the losers in Z^1 so that the losers are no worse off than in Z^1, then Z^2 is a preferable state even if no compensation were actually paid. In response to the Hicks-Kaldor thesis, Scitovsky[9] has pointed out that both Z^1 and Z^2 could be shown to be socially preferable states such that the gainers in Z^2 could compensate the losers in transformations from Z^1 to Z^2 and from Z^2 to Z^1.

But it is obvious that the situations described by compensation theorists are highly contrived and unrealistic. More important, these models entail the formulation of interpersonal comparisons of utility, which are explicitly ruled out by the Pareto criterion and Arrow's theorem. Under these conditions, improvements are nothing other than efficiency adjustments on the Pareto exchange contour.

But if the theorist as a welfare economist must necessarily make recommendations concerning the social costs and benefits of economic programs, and offer prescriptions about equity and the merits of distribution, then, clearly, welfare economics is a normative discipline. In fact, the general view is that the term "welfare economics" itself implies an economics concerned with ethical criteria. Thus, any notion such as "positive welfare economics" would be a contradiction in terms.[10] It is obvious, though, that the neoclassical system is a logically consistent whole in the sense that the analytical framework of the neoclassical microeconomics theory implies the Paretian welfare economics theory. Questions on distribution, equity, and so on, as the theory is understood, are not at all entailed by it.

It is on this basis that one disagrees with the orthodox notion emphasized by Blaug that welfare economics belongs unequivocably to the area of normative economics as distinct from positive economics. On the other hand, the view expressed by Hennipman that there is no qualitative distinction between positive economics and Paretian welfare economics is one that is logically correct. I qualify this claim, however, with the thesis that while Hennipman views the whole corpus of neoclassical theory as objective in its claims, the research in this study demonstrates that neoclassical theory as a whole (i.e., both welfare and positive economics) is normative in content.

Value Judgments, Neoclassical Microeconomic Theory, and the Paretian Model

In the earlier chapters of this text, I argued that on account of the evidently normative role played by the postulate of rationality and its attendant axioms, neoclassical microeconomics was necessarily compromised in its scientific pretensions. I reiterate that the same may be said for the Paretian welfare economics model. First of all, the axiomatic structure of Paretian welfare economics is structurally identical with that of the positive microeconomics model, as the formulation of it demonstrates. The sole difference between the axiomatic assumptions of both models is that Paretian welfare economics attempts to formulate a social welfare function from a set of individual rankings.

Proof of the claim that there is an evident coincidence in terms of axiomatic structure between welfare economics and the microeconomic model is obtained from the formulation of the basic theorem of theoretical welfare economics:[11] (i) Every competitive equilibrium is a Pareto optimum, and (ii) every Pareto optimum is an equilibrium state of the competitive economy. The two parts of this basic theorem have been called the "direct theorem" and the "converse theorem." Surely, if both theories (i.e., the neoclassical "positive" theory and its normative counterpart) were qualitatively different, then the above-mentioned basic theorem could not have been logically defensible.

Given that the major aim of this study is to show that at its very foundations neoclassical microeconomic theory is founded on evidently normative assumptions, let me now prove, in the light of the above, my thesis. Call the neoclassical microeconomic theory A, the general equilibrium theory B, and the Paretian welfare economics theory C. It has been demonstrated above that A is a normative system and that A entails B. Now call the combination of A and B, Z. Clearly, if Z implies C and Z is normative, then C must also be normative.

The above discussion, no doubt, runs counter to the belief held by most neoclassical theorists that there is a justifiable theoretical distinction between normative and positive economic theory. But they would certainly be at loss to justify the basic theorem of theoretical welfare economics. If the neoclassical theorist accepts the claim that welfare economic theory is normative and also accepts the basic theorem of welfare economics, then he or she must also claim that neoclassical microeconomic theory is derived from normative assumptions. Elementary logic states that factual propositions cannot be logically derived from normative ones. The converse is also true. The alternative would be to accept the notion that neoclassical welfare economics is cognitive in content and a theoretical device concerned only with efficiency. But this view could be maintained only if "positive" theory were shown to be non-normative. It has been demonstrated that this is not the case.

Normative Constraints and the Neoclassical Welfare Economics Model

This discussion has amply shown that the whole structure of neoclassical theory is normative in nature. This would mean that the question of economic methodology would entail value judgmental considerations entailing serious questions about ethical decision making, equity, justice, and so on. It is in this context that neoclassical welfare economics is linked with the recent contemporary ethical research of Rawls,[12] Nozick,[13] Sen,[14] and others.

It should be noted though that most of this research is highly hypothetical in nature and does not offer any real improvement in answering the important normative questions of economic decision making. A further problem is that the above-mentioned theorists take the neoclassical economics framework for granted and do not raise questions about the empirical structures on which the neoclassical framework is founded. Thus, questions about the fundamental role of capital as a factor of production and its relation to other factors of production are rarely discussed by those theorists who raise questions about the problems of distribution, equity, and the like.

For example, Rawls and Nozick emphasize the importance of liberty in their attempts to formulate the criteria of an acceptable social collectivity. Rawls's two principles of justice and Nozick's defense of the free-exchange economic structure are all compatible with the neoclassical framework. Although Rawls's "veil of ignorance" theoretical assumption may be viewed as guaranteeing equality in terms of individual liberties, his second principle of justice is indeed compatible with the Pareto criterion. Recall that Rawls's second principle of justice states that inequalities in power, wealth, income, and so on are acceptable if they exist for the benefit of the worst-off members of society. Clearly, the finality of this proposition corresponds with an equilibrium in terms of Pareto efficiency.

Nozick, on the other hand, advocates what is described as a historical procedural approach to the problem of economic decision making within the social collectivity. This theory of social justice is founded on principles determined by the way in which assets are acquired, transferred, and past injustices rectified. Nozick's thesis is theoretically compatible with that of Rawls in that he emphasizes liberty and contract-based rights as the basis for defining a collectivity of economic agents. But the weakness of this thesis is that it does not contain any mechanism whereby rights may be ranked, thereby allowing the possibility of the redistribution of assets. The importance of the rights approach of Rawls and Nozick is that once one makes the assumption that any discussion of theory formulation in economics must be founded on normative assumptions, then the construction of any practical economic model must take into consideration the question of rights.

I state this because economic decision making and the theories constructed to explain it are never constructed in a social vacuum. Theory formulation in economics must implicitly take into consideration the notion of the social framework, the embodiment society's ethics, laws, and so on. But ethics and laws entail decision making within the context of constraints and freedoms. In other words, they entail rights that may be formulated and defended mainly on the basis of *a priori* principles.

Note finally that the role of theorists such as Rawls and Nozick in the discussion concerning welfare economics is to seek to justify the validity of the Paretian framework against those who question the foundations of the neoclassical mechanism as a normative guarantor of a market system in which the most important methodological criterion is efficiency. And it is efficiency that is viewed as absolutely more important from the theoretical point of view than equity, social justice, and other concerns.

9
Alternative Methodologies

In the previous chapters, I have argued that the now dominant neoclassical theory is deficient in its claims to scientific status given the normative role played by the foundational postulate of rationality. This critique applied not only to "positive" neoclassical theory but also to orthodox Paretian welfare economics in the context where it is viewed as a component of positive theory. Having engaged in a systematic critique of neoclassical theory, the main purpose of this chapter is to discuss theories of economic decision making alternative to the neoclassical model, then to examine the viability of a synthetic theory founded on elements culled from a variety of social science disciplines. This is only fitting given the critical evaluations that constitute the major portion of this text.

The approach I take in this chapter is to discuss in cursory fashion some of the more important theories of economic analysis that have grown in popularity in recent years given the reservations that many theorists express toward the dominant neoclassical school. In the context of the discussion, it should be understood that the alternative methodologies I choose to discuss are deficient in the sense that they do not properly and thoroughly examine the foundational structure of the neoclassical theory as a component of their analysis. As a result, their proponents implicitly structure their particular methodologies and theories on questionable aspects of the neoclassical theory. The research schools I propose to discuss (the list is not exhaustive) are the Austrian model, institutionalism, political economy, post-Keynesianism, and Marxian socialism.

The Austrian School

The theoretical basis for the contemporary Austrian theory derives from the original work in economic theorizing begun decades ago in Vienna under the leadership of Carl Menger. The ensuing research of Ludwig Von Mises and Friedrich Hayek further helped lay the foundations for what might now be called the Austrian paradigm.

Although the Austrian school of economics is not of recent origin, it attracted little attention in those research centers where neoclassical economics held sway. The reason for the renewed interest derives, no doubt, from the methodological problems that now beset the neoclassical theory. The Austrian critique of contemporary neoclassical economics is founded on the claim that the latter places too much emphasis on a static end-state concept of equilibrium and too little concern with the actual dynamics of economic decision making. Austrian theorists also argue that the neoclassical claim that rational choice may be objectively definable is false; they argue rather that rationality as the basis of human choice making is a subjective and *a priori* concept.[1]

James Buchanan, though critical of Mises's overly general application of the concept of subject economic decision making to all modes of human behavior, does support the key Austrian thesis that deliberate choice making is subjective and "not amenable to scientific explanation."[2] Buchanan's confidence in his claims derives, no doubt, from what he describes as the failures of the predictive science of economics, that is, the orthodox neoclassical model.[3] However, despite Buchanan's methodological modifications, the central claim in the Austrian thesis is that there can be an *a priori* science of human choice that is both logically valid and empirically based. Historians of ideas would, no doubt, perceive a Kantian influence in this claim. The result of this approach is that the subject matter of economics for the Austrian theorist is the subjective thoughts or intentions of individual agents. It is argued that attempts at objective analysis from a positivist perspective is impossible given the indeterminacy of human action as a function of the thoughts of individuals. Proof of this thesis is claimed to issue from the fact that when all individual actions are coordinated within society, the result could be quite different from the intentions of the agents themselves.

The importance of the Austrian thesis is based on the fact that it highlights the difficulties involved in formulating a predictive science of human choice. It also points out that the important notion of utility is intrinsically a subjectivist notion properly homologous to nonempirical terms such as "purpose," "motivation," "belief," and so on. Paradoxically, too, the model demonstrates that the orthodox neoclassical assumption of rationality is not an empirically determinate concept but one assumed *a priori*. The main difference between both theories in reality is that the neoclassical model claims to explain and predict the empirical choices of economic agents, while the Austrian model argues against this possibility.

The weakness of the Austrian model, on the other hand, can be explained in terms of its incapacity to propose any testable or determinate claims about the economic processes in society. Of course, one can, as Kirzner[4] does, comment on the economic process in fairly general ways, but the Austrian theory allows little scope to introduce specificity into economic analysis. Yet to propose otherwise would be to contradict one of the Austrian theory's important assumptions—the absolute priority it affords to the concept of freedom. And this concept, no doubt, serves as a necessary condition for free-market economic systems. But free-market economic systems whose dynamic paths are indeterminate by definition can lead only to anarchy. Given these assumptions, it is difficult to discern how any theorist could hope to formulate empirically consensual propositions.

Institutionalism

Contemporary institutionalism represents another influential rival to the dominant neoclassical model. The research efforts of institutionalists like Gunnar Myrdal and J. K. Galbraith are well known. Orthodox institutionalism claims that contemporary neoclassical theory, with its structured models of analysis in which only a finite set of variables are posited, yields results that bear minimal correspondence to reality. The institutionalist would argue that a proper analysis of economic behavior should entail the formulation of models that include not only purely economic data but also sociological and historical information. In fact, the influential institutionalist Gunnar Myrdal would argue that the institutionalist model is much more compatible with the original goals and functions of classical political economy. And for Myrdal, the term "economics" is short-hand for "political economy," with its open acknowledgment that valuational assumptions are a necessary component of economic analysis.[5]

Other institutionalists argue that the research goal of the institutionalist is to seek understanding (rather than explanation) of economic behavior in terms of "pattern models" grounded in institutional and cultural frameworks.[6] In short, the approach of the institutionalist is interdisciplinary and more sociologically descriptive than that of the orthodox neoclassical theorist. The institutionalist's research goals are to seek forecasts rather than predictions in terms of some predictive model, as the neoclassical theorist is wont to do.

The possible neoclassical response to the institutionalist program would be that it does not possess a well-formulated theoretical structure according to which proper empirical analysis could be carried out.[7] The lack of such, it would be assumed, makes it difficult for the institutionalist to establish objective scientific analyses of economic behavior. Although the neoclassicist would be justified in offering this criticism, I am inclined to believe that institutionalism does have some positive features in that it is better equipped to offer more useful causal analyses of long-run macroeconomic trends. It is also better equipped to offer more complete analyses of economic growth and development trends in societies where historical and cultural variables differ from those where the neoclassical model is dominant.

On the basis of the above discussion, I believe that the conditions are appropriate for the theorist of institutionalism to establish testable theoretical models that would serve as the basis for their recommendations. These testable theoretical models would be founded, most likely, on the synthesis of a number of positions that in time would be expressed as a general theory. Yet although the institutionalist research program appears to be an effective antidote to the evidently irrelevant modeling of the neoclassicists, its weaknesses are that it does not yet possess a structurally firm model of microeconomic behavior and it seems to lack clear prescriptive programs that could be effective as policy tools for problems both at the microeconomic and macroeconomic levels.

Political Economy

Political economy theory is a growing school of thought that presents itself as an alternative to neoclassical orthodoxy. The political economists share some affinity with the institutionalists in that their analytical models are founded not only on purely economic factors but also on political considerations. But contemporary political economy differs from institutionalism in that its adherents frame their analyses and prescriptions from well-structured theoretical frameworks. Within the research program of political economy, three methodological orientations are evident: Marxian political economy, neoclassical political economy, and public-choice political economy. Marxian political economy is, perhaps, the most easily recognized of the above-mentioned approaches since the traditional Marxian analysis of economic behavior has always sought to establish a causative relationship between economics and the political behavior of groups or classes in society. Marxian political economy has also sought to conflate into one theory of analysis both an empirical study of economic behavior in society and a normative evaluation of the theory's empirical observations. According to the theory, the relationship between the owners of capital and those who supply only their labor is an exploitative one and ought to be struggled against. At the same time, orthodox Marxian political economy claims that the scientific laws of capitalist motion predict the collapse of capitalism as an economic system. The theory also predicts that the next great stage in the evolutionary development of economic systems is that of socialism/communism. Thus, one witnesses in Marxian political economy an interesting theoretical merger of value and objectivity. This means that although the Marxian political economist would claim that the Marxian analysis of the capitalist economic system is scientific, his prescriptions are openly value judgmental.

In this connection, Marxian political economy views contemporary neoclassical theory in strictly ideological terms: neoclassical theorists are viewed as conscious or unconscious defenders of the political status quo of the capitalist economic system. Marxian political economy would also claim that the utility or profit-maximizing choice paths of economic man, as formulated by neoclassical theory, represent a caricature of actual economic behavior and a theoretical apology for unbridled consumption and profit making as the basis of human behavior.

Neoclassical political economy as a branch of contemporary political economy is, perhaps, the least radical of the above-mentioned group. The reason for this is that the problems and methodology of neoclassical political economy are almost identical with those of neoclassical Paretian welfare economics, with its emphasis on the analysis of such concepts as efficiency, equity, cost-benefit analysis, and so on. I have already discussed neoclassical welfare economics and its relationship to orthodox positive neoclassical theory; there is, therefore, no need for further comment except to state that of the three approaches to political economy discussed in this section the neoclassical model is the least radical.

Public choice political economy has recently attracted much interest given the research efforts of theorists such as Buchanan and Tullock.[8] Public choice political economy could be defined as that research school which attempts to explain the behavior of political institutions, government officials, bureaucracies, and so on by appeal to the traditional techniques of neoclassical economics. An important assumption in public-choice theory is that public officials actual and potential seek to maximize utility in the political marketplace in ways similar to those of the entrepreneur in orthodox neoclassical theory. But there is some theoretical disagreement as to the kind of "marketplace" in which buyers and sellers trade. Buchanan and Tullock would argue that because of the monopolylike features of government institutions and bureaucracies, great inefficiencies exist in the public-choice arena. The key general point in this approach to the political economy of public choice is the generalizing of the neoclassical notion that all agents seek to maximize their own personal utility functions whether in the economic or political arenas.

Until the time of Buchanan and Tullock, the quasi-economic analysis of political decision making was greatly influenced by the research of theorists such as Downs,[9] who argued for a demand model of political decision making: the government simply attended to the inefficiencies of the market system for the benefits of voters. This approach to the politics of economic decision making could be viewed as the theoretical aftermath of Keynesian theory. More recently, given the evident theoretical clamor in some quarters for "less government," and the increasing popularity of "classical liberalism" as a political philosophy, it is fitting that the supply models of Buchanan, Tullock, and others would seek to establish the justification for "less government" by demonstrating in economic terms that the fiscal activities of government as a supplement to the market economy produced inefficiencies. In this context, it is useful to discern in public-choice political economy a theoretical connection between the Austrian economics school and the current interest in liberalism as a political philosophy with the concomitant thesis of "minimal government." It should be noted, too, that public-choice political economy is methodologically akin to orthodox neoclassical theory in that it maintains the traditional distinction between positive and normative theory.[10]

In sum, it would seem that from this brief discussion of trends in contemporary political economy, it is evident that it is Marxian political economy that seriously seeks to formulate a model radically at variance with the neoclassical model. Public-choice and neoclassical political economy both seek to analyze economic phenomena as they are influenced by political considerations but from within the framework of orthodox neoclassical theory.

The Post-Keynesian Model[11]

Post-Keynesian theory has gained attention in recent years, given the persistent problems of unemployment and inflation that have intermittently plagued modern industrial society. Post-Keynesian theory could be viewed as a synthesis of ideas from both Keynesians and neo-Keynesians. The basic critical claim of the post-Keynesian school is that the neoclassical general equilibrium theory does not adequately describe actual empirical events in the general economy. It is argued that the formal neoclassical structure produces only static models with no mechanism for dynamic, empirically grounded models, which could take into account the vagaries of the actual economic world—the "rational expectations" theoretical appendage to the neoclassical theory notwithstanding.

Thus, the fundamental tenet of the post-Keynesian theory is that economic analysis is significant only if empirically grounded. This is in contrast to the purely formal general equilibrium model of the neoclassical theory, which claims that at "equilibrium" all markets are cleared. On the other hand, post-Keynesian theory has maintained, on the basis of empirical analysis, that there could be equilibrium at which not all markets were cleared. The post-Keynesians are also noted for their attempts at establishing holistic models of analysis that incorporate both political and economic variables.

I am inclined to believe that the post-Keynesian program is relatively promising but still lacking in maturity. Its major weaknesses are that it has not yet been able to establish both a well-structured critical theory of neoclassical economics and a coherent structured theory of economic analysis.

Marxian Socialism

Perhaps no alternative theory to neoclassical economics has generated more controversy than theories of socialism founded on the ideas of the nineteenth-century theorist Karl Marx. The main thrust of Marx's research was to synthesize the political economic ideas of Smith, Ricardo, and Malthus with the political philosophical ideas of thinkers such as Rousseau, Proudhon, and Hegel. Yet the most important aspect of Marx's research is the critique of the dynamics of nineteenth-century capitalism in his magnum opus, *Das Kapital*. Marx attempted to show that capitalism as an economic system was an inherently unstable system doomed to collapse because of the imbalanced relationship between the owners of capital and those employed by capital. Given the role that industrial workers play in the dynamics of capitalism, Marx argued that the transition from capitalism to socialism would take place in the relatively advanced industrialized societies. History has not borne this prediction out, nor has it shown that workers in advanced industrialized societies are inclined to revolutionary activity in the Marxian sense.

On the contrary, the so-called socialist revolutions took place in the then predominantly agricultural societies of the Soviet Union and China. Intellectual supporters of Marxian socialism argue that socialism, as practiced in the Soviet Union and China, offers a better economic alternative than capitalism because of the nonexistence of a class of private capital holders and a much more equitable distribution of wealth and income. The claim here is a normative claim derived theoretically from Marx's attempts to analyze the capitalist economy scientifically. Orthodox contemporary socialist economic thought does not pay much attention to microeconomic theory and prefers to subordinate economic analysis to macroeconomic theories of state planning. In this connection, the state as a political entity is seen to possess great influence. Thus, the orthodox socialist model of economics frames economic analysis in terms of what the state planners view as the needs (as opposed to wants) of the populace. The result of this, of course, is generally some form of rationing and supply quotas. Obviously, this model is radically divergent from the demand-supply driven models of the market economies.

For purposes of this analysis, it is indeed important to note that orthodox socialist economic theory is regarded as scientific by its adherents. Marx claimed to have discovered the laws of motion of capitalism, and his thesis that capitalism as an economic system would eventually fail was viewed by him and his intellectual followers as having predictive validity. Yet at the same time, there was an evident moral and exhortative character to the Marxian justifications for socialism. And there is a curious historical irony in the way in which socialist economic theory is justified as scientific in those societies which claim to be socialist. Given the structural subordination of market mechanisms to the control of the state planning instrument, the problem of the scientific nature of economic decision making in the sense of explanation, prediction, and control is much resolved since the choices of agents and plant managers are strictly determined by the constraints of the state plan. There is no room for decision uncertainties if the consumer is required to purchase only those items produced by the state firm. But even in a state-controlled economy, there are still instances of errors in prediction given that shortages in supply occur with some frequency.[12]

Given the above discussion, the important question is how

plausible is the socialist economics thesis as an alternative to the neoclassical model. I am inclined to believe that the socialist economics thesis addresses issues that the neoclassical economics program neglects. The socialist model has rejected the notion that economics as a scientific discipline is meaningful without methodological concern for its instrumentality in the increasing of human welfare. The methodological distinction that neoclassical economics makes between positive and normative economics is not countenanced by the socialist model. However, the major methodological problem with the socialist model is that much of its theory is regarded as unassailable by most of its adherents. Empirical events that seem to falsify the claims of the theory are disregarded by appeals to *ad hoc* propositions and *ceteris paribus* clauses. I admit, of course, that it resembles the scientific career of the neoclassical theory in this instance.

Yet the socialist alternative has been subject to criticism on account of its highly constrained analysis of economic decision making leading to the restrictions on human choice and freedom. There is some truth in these claims. It is highly interesting, though, that while Marx's original thesis was to show how workers were unfree in the sense of not being owners of the capital with which they worked, contemporary socialist theory does not supply the logical answer to the problem. The logical solution to Marx's recognition of the relationship between labor and capital is that workers own and manage capital in very much the same way in which the capitalist owns and manages capital. The result of this would be a society of decentralized economic units in which the political power of the state would be much reduced. Yet this was not Marx's solution. He argued instead (though in rather sketchy fashion) that in the aftermath of the proletarian revolution, the socialist state, guided by the communist party, would protect and control the means of production.[13]

The result of this was that the socialist revolutions for which Marx argued led to the transformation of capitalist societies not led by workers (the proletariat) but by intellectuals and technocrats, all members of communist parties. It is for this reason that most socialist states were characterized by a preponderance of state power in the hands of communist parties led by socialist intellectuals and bureaucrats. Although the members of the revolutionary parties in the orthodox communist states do not actually own capital, they effectively control it, given the extent of their political powers.

The well-known historical product of this has been the proliferation of state bureaucracies that controlled the economic, cultural, and political life of the citizens of the communist state. This excessive control of the life of the individual produced not the economic and political freedom that Marxism promised, but in many instances a stifling political environment and lack of intellectual freedom.

Recent events in the Soviet Union and Eastern Europe offer empirical proof that although Marx's critique of capitalism was sound, the communist alternative is being rejected by those who have had first hand experience of its practice. Yet because of the social problems engendered by market capitalism as justified by neoclassical economic theory, the socialist economic model, as inspired by Marxian economic thought, still has its adherents. And there are theorists who are still seeking to formulate a theory founded on the conciliation of market economics and socialist economic theory.[14]

The Question of Methodology in Retrospect

Economists in general do not spend much time ruminating on the methodology of their discipline, the reason being that they have not been trained in that direction. They prefer to interpret economic phenomena from the vantage point of their respective research paradigms. But given the questions increasingly raised about the effectiveness of the discipline in analyzing economic phenomena, a growing number of theorists are now turning to questions of methodology. The confident foundational work of Jevons, Walras, and Pareto, later embellished by Friedman, Samuelson, and Hicks, is now being examined from the standpoint of methodologists of science such as Popper, Kuhn, and Lakatos. Mark Blaug's *Methodology of Economics* and Bruce Caldwell's *Beyond Positivism* are relevant examples of this trend.

Blaug, expressing reservations about the state of current research in economics, prescribes that economists take seriously the idea of theory falsifiability.[15] His methodology is influenced, no doubt, by Popper's ideas on scientific research. Caldwell, similarly concerned about the scientific status of contemporary economics, advocates methodological pluralism as the best way forward. Caldwell argues that "no single optimal methodology is discoverable,"[16] and that science thrives better in a free research environment than otherwise.

But positivism still seems to exercise much influence on the methodology of economics, given that the belief is still maintained that a positive science of economics is theoretically possible. The implicit assumption is that a necessary and sufficient condition for the potential scientific status of a research discipline is the quantitative expression of its claims. The fundamental questions about the nature of human decision making persistently posed by phenomenologists are ignored by those who believe that such questions are rendered irrelevant by the quantitative propositions of neoclassical theory and econometrics. But it is only an idealist theorist who will not recognize that neoclassical economics and econometrics do not deliver on their explanatory and predictive promises.

As mentioned above, the theoretical puzzles of contemporary economics will remain unsolved until methodologists pose again questions concerning (a) the nature of human behavior, and (b) the purpose of economic decision making. Answers to both concerns would seem to entail a methodology founded on the idea of facts in the service of value. The point is that economics can only be a policy science. It could not be otherwise, since economic decision making entails the effecting of choices according to normative programs of optimality. The two central commandments of neoclassical theory issuing from the postulate of rationality are "maximize utility," and "maximize profits." And the theorist who constructs a program advocating these two central principles is, no doubt, advocating policy. Of course, the neoclassical theorist, trained to reify the concepts of rationality and the maximization of utility and profits, would tend to disagree with this view, but I have argued that economics is really about optimal decision making according to some principle of correct action, that is, some normative principle of choice. The question of economics reduces then to the issue of formulating consensually acceptable theories of optimal decision making. I shall explore this idea in the following chapter.

10
A Theory of Optimal Decision Making

The previous chapters in this text have been concerned primarily with a critical analysis of neoclassical economic theory in terms of its claims to scientific status. In this final chapter, I propose to offer an exploratory program for a theory of economic decision making founded on the assumption that human decision making necessarily derives from normative principles. This would mean that the formulation of any theory of economic decision making would entail analyses of both means and ends of decisions. But it is evident that the means and ends that individuals choose when they appeal to their decision schedules are subject to technical constraints. Thus, it is possible that an individual choose means and ends that are technically unattainable. Quite obviously, the formulation of decision schedules entails questions of optimality. The problem of formulating a theory of economic decision making would involve, therefore, posing questions of optimal decision making both for the individual and for the sum of individuals in society, that is, the social collectivity. Granted that all individuals would be concerned to effect optimal choices in terms of means and ends subject to technical constraints, the formulation of normative theories of economic decision making reduces to the formulation of optimal theories of choice.

Given the above, the rest of this chapter will be concerned to examine the possibilities of establishing criteria necessary for the formulation of a theory of optimal economic decision making for individuals within the constraints of society. I shall proceed in the following way. I shall base my exploratory theory of optimal economic decision making on the thesis that human beings are especially disposed to effect a multiplicity of choices. This is the basis for the great significance that political philosophers ascribe to freedom in human affairs. Given that the choices individuals make are effected within the confines of society, questions on the limitations of individual freedom arise. Thus, the formulation of a theory of optimal economic decision making must first take into consideration the constraints imposed on individual choice making by society. A resolution of the problem of constraints would entail, therefore, discussion of the issues of liberties and rights within society.

The formulation of a theory of optimal economic decision making would involve, therefore, the exploration of schedules of choice that seek the maximization of possibilities to effect economic choices within the constraints of society. The problem with the neoclassical welfare economic theory in this regard is that it does not set the social parameters or boundary conditions within which individual choice making is maximized. The simple assumption that individuals are purely self-interested in their choice patterns does not do justice to the nature of economic decision making within the social structures of the market economies. Thus, before presenting a theory of optimal decision making, I shall first point out the deficiencies of neoclassical welfare economics as a theory of normative decision making for society as a whole.

I shall then propose a brief exploratory theory of both the micro- and macroeconomy. In the discussion to follow, it would be apparent that an attempt is being made to view optimal economic decision making as being bound up with ideas of political philosophy.

Paretian Welfare Economics and Its Limitations

Neoclassical welfare economics is viewed as that branch of neoclassical economics which is concerned with the welfare of society as a whole. It is founded on the assumption of the selfishness of individuals and the derived axioms of rational choice that constitute the foundations of neoclassical microeconomic theory. In terms of optimal decision making, the maximum that this theory entails is what is known as Pareto optimality. A social state is Pareto optimal if and only if any further trade of assets among individuals would leave at least one individual worse off.[1] Yet there is some question as to how optimal social states could be attainable given the assumed proscription against interpersonal comparisons of utility. The basis for this, one recalls, is the long-standing argument against the possibility of measuring utility.

Paradoxically, the theory itself, derivative from classical utilitarianism, is quite incompatible with classical liberal political theory. While neoclassical economics theory saw fit to embrace Bentham's measurable utilitarianism, liberal political theory, on the other hand, preferred to endorse a Millian-type liberalism with its emphasis on the negative liberties.[2] If liberty were viewed as an economic good—in many societies, on account of its scarcity, it is virtually viewed as such—then Pareto optimality would be quite compatible with politically authoritarian societies in which the distribution of freedoms is unequal. In fact, proof of the weakness of Paretian welfare economics as a theory for optimal decision making is that no Rawlsian individual would choose such a mechanism as a means of optimizing individual welfare. My approach is radically different from that of the neoclassicist theorist in that it is founded on the assumption of the individual not as a socially atomistic self-interested utility maximizer but as an individual to whom rights to a maximum set of liberties have been guaranteed.

In this regard, the appeal by the contractarian and libertarian public-choice theorist to the mechanisms of Pareto optimality for the resolution of social conflict problems is rather inconsistent given that social-contract theory necessarily requires that basic incommensurable rights be established *a priori* in order to establish the contours of the social framework. Thus, contractarian Buchanan's[3] hesitation to anchor the libertarian social contract with specific constitutional rights no doubt leaves the door open for compatibility with neoclassical Pareto optimizations. The point is that there can be no genuine contractarian approach for purposes of theory construction unless specific rules are first established. In fact, it is just this point that underscores the incompatibility of neoclassical Pareto welfare economics and a rights-based contract theory of liberalism. And it is only the latter that is equipped to specify the rules that would determine the contours of the social collectivity.

Decision Making in a Social Context

Although economic decision making generally takes place within societies among its members, consider first a single individual social collectivity. Under such circumstances, the individual's decision schedule would be constrained only in terms of the economic resources available, that is, his or her own labor, technology, and natural resources. Individuals would experience no social constraints whatsoever in terms of their thoughts and their vocal or public expression, coercion by other individuals, freedom of movement, protection of their own resources, and so on. Granted the relative complexity of the human brain, the individual would be subject to the experience of a continuing series of thoughts, some of which would be sought to translate into acts. It is the ordered sequence of such thoughts that constitute the individual's decision schedule.

I elaborate on the above point by pointing out that the characteristic that evidently differentiates the human species from other biological forms is its capacity for great cerebral activity, which is translated into a multiplicity of thoughts and decisions. In other words, human beings are constantly seeking self-expression in terms of conscious thoughts and decisions. It is just this characteristic, I believe, that explains the great premium that individuals place on freedom. For this reason, restrictions on the latter are always met with resistance.

But the pressures of evolutionary biology have ensured that human beings, like so many other biological species, exist as members of social groups. Consider, for example, the obvious social function of language. Yet whatever advantages individuals derive from membership in social collectivities are gained at the cost of restrictions on individual freedoms—for ideally, each individual sees himself or herself entitled to as much freedom as any other individual within society. Such are the constraints imposed on unrestricted freedom within society that it is effectively transformed into a scarce commodity over which individuals compete. If such competition is given free rein, what eventuates is a state of social anarchy—the proverbial state of nature used as a reference point by political philosophers such as Hobbes and Locke. This is the sociological origin for the existence of social codes and ethical principles in every known society. It is also the basis in more recent times for the formulation of social-contract theories formulated by theorists such as Rawls and Nozick.

Yet the idea of freedom entails the totality of man's social life, hence its relation to the concepts of economic and social justice. One question is whether individuals should be free to appropriate as much of society's economic resources as possible regardless of the size of individual shares. In this regard, the theoretical efforts of welfare economists such as Bergson, Little, and Arrow are instructive.

Thus, on the one hand (in the case of the "rights to liberties" social-contract theorists), there are attempts not only to maximize liberties in society but also to establish tolerable equilibria of such. On the other hand (in the case of the welfare economists), the goal is not only to maximize a society's economic resources but also to establish just equilibria of its shares. But as pointed out above, the research efforts of the welfare economists are compromised by the theoretical limitations imposed by the principle of Pareto optimality and the lack of concern on the part of the neoclassical economist for issues relating to ethics and rights.[4] And in the case of the orthodox social-contract theorist only the negative liberties[5] (especially the political liberties) are pertinent to any discussion on rights. There is little concern to extend the discussion on rights to the economic sphere. It is assumed that decision making in the area of political choice is qualitatively distinct from that of economics. Incidentally, it is this putative distinction between the political and economic lives of individuals that has led to the notion in the liberal democracies that political freedoms are qualitatively superior and distinct from issues of economic welfare.[6] Similarly, in statist societies the authorities often claim that curbs on political liberty are necessary to ensure economic justice for the individual.

The central point of my argument, though, is the claim that if the political liberties (i.e., the orthodox negative liberties) should constitute basic rights in any adequate social contract, then an optimal exercise of these rights presupposes certain rights in economic decision making.[7] I shall elaborate on this point in the following section.

Rights, Social Justice, and Human Capital

In this section, I want to argue that any general theory of economic decision making must take into consideration those normative questions relevant to the structure of society in terms of individual rights and social justice. The issues concerning social justice and rights that neoclassical economic theory relegates to welfare economics and political philosophy would now be viewed as important in the formulation of general theories of economic decision making.

Perhaps a key question concerning the structure of any society is that concerning the extent of liberties that could be extended to its members. Indeed, the theoretical concerns of Rawls,[8] Nozick,[9] Buchanan[10] and others attest to this. The consensus has been that individual liberty is a primary good, necessarily restricted for each individual within the confines of society. In practice, though, such liberties are limited only to the traditional set of negative liberties.[11]

Arguments are then proposed claiming that within society the individual should have rights to such liberties. However, such rights are not extended to the important positive right to engage in income-generating work. In fact, I want to believe that it is the ability to create exchange value within society that guarantees the stock of traditional negative liberties. Rawls, for example, establishes a firm distinction between liberty and economic goods.[12] According to this approach, liberty and economic goods are tradable commodities within the developmental path of the social collectivity. But it is the human activity of work that is the generator of wealth, which, in turn, facilitates the traditional liberties. Evidently, the individual not engaged in exchange value generating work—reference is not made here to individuals who have inherited or previously acquired wealth—would experience severe practical restrictions on his or her rights to the traditional liberties. Furthermore, if one views liberty as entailing the idea of "forms of self-actualization"[13] and if one views subjectively meaningful work as also leading to self-actualization, then clearly "work" and "liberty" are cognate terms.

There are good grounds for arguing that the continued existence of human societies over time derives from the economic advantages gained from cooperation in the form of the division of labor in society. In their continuing quest to control the forces of nature for survival, human beings have sought to employ whatever technology that was available. It was the available technology conjoined with human activity that constituted work. And the purpose of work was simply to attain human ends. But again, it is work in the form of technology manufacture and its improvement that further facilitates the activity of work. Thus, throughout the course of history as technology and modes of work improved in efficiency, human beings have been able to realize an increasing number of goals. Now, if the realizing of one's goals constitutes the exercise of freedom, then human freedom may be seen to have increased historically in direct proportion to the effectiveness of work. It would seem, then, that modern humans, given their increasing capacity for choice making, are potentially more free than their ancestors. Thus, it appears that human freedom is a direct product of work. To put it more concretely: the individual in the modern exchange economy feels most free when able to actualize himself or herself through fulfilling work, the remuneration of which allows the earning of exchange value necessary for the effecting of economic choices in the marketplace.

It seems, therefore, that if one approaches the problem of establishing criteria for an optimal social collectivity from the contractarian point of view, then it would seem that one of the conditions for membership ought to be the right to productive labor as a guarantee of freedom. In this way, the great significance and prior conditionality assigned to the right to liberty by contractarians like Rawls and Nozick can be meaningfully operationalized in society under these economic conditions. Consider the fact that in the real world, defense of one's stock of negative liberties requires potential appeal to one's economic resources in the form of obtaining legal counsel and so on. But if effective work in society presupposes training and education, it is obvious then that the right to work presupposes education and training. The individual entering the modern market economy, ignorant of his or her prospective status therein, would do well to view the possibility of genuine liberty guaranteed by a prior education and productive work.

In fact, modern industrial society partially endorses the above claim by the legal requirement that all individuals invest in their own human capital by means of public education. Compulsory childhood education in modern society is a well-known phenomenon, so it certainly appears paradoxical that the modern liberal democracy would enforce investment in personal human capital and not require that the individual have the right to productive labor given its evident role as a precondition for enhanced liberties. Proof of this claim is the fact that individuals who are involuntarily employed view themselves as having lost freedom, in the sense of feeling thwarted from realizing legitimate choices. Furthermore, individuals who are involuntarily employed for periods of time tend to seek to migrate from their own societies. In fact, one might note that in the course of history economic refugees have greatly outnumbered political refugees.

But most rights theorists, whether of libertarian outlook or otherwise, rarely extend the idea of rights beyond the traditional negative liberties. It appears, though, that the restricting of rights to the basic negative liberties assumes a very simple artificial society, unlike the market economies or statist societies of the modern world. I imagine that the idea behind the rights to negative liberties is that of the autonomous individual, but as argued above, liberty for the individual in the market economy presupposes a set of economic rights.[14] In this regard, therefore, one could take issue with political philosopher Rawls's highly influential principles of justice. Rawls's first principle states that "each person is to have an equal right to the most extensive basic liberty compatible with a similar liberty for others," but the second principle concerns only social and economic arrangements. There is no reference to rights here. But in the context of society, the obvious question that arises concerning the first principle is "liberty for what ends?" The orthodox answer to this question would focus on the basic stock of political liberties already formulated by liberal theory. Yet Rawls's second principle concerns the concrete and important issues of social life, that is, matters of economic deserts and social ranking. There appears to be a logical gap between the first and second principles that most authors fail to recognize.

If Rawls's concept of equal liberty could be qualified to mean rights to liberty to acquire those means which would lead to the ends of the optimization of individual liberties within society, then his two principles of justice could be made compatible. In fact, this qualification reinforces the idea of the rights liberal theorists usually defend as being necessary for an adequate contract theory. Thus, the illiterate and ill-informed individual is less able to optimize his or her right to freedom of expression as the literate and informed. Similarly, the uneducated and ill-trained individual is unable to optimize his or her implicit right to freedom of choice in consumption and choice in occupation. According to the main point of this chapter, those means which could lead to the ends of the optimization of individual liberties entail prior rights to education, training, and self-actualization in work. Thus, given a more grounded first principle, then Rawls's second principle could be reformulated to mean that differences in individual income may result in advantages for society as a whole. Without a qualified first principle, Rawls's second principle (I refer specifically here to the difference principle) is indeed open to criticism since it is unlikely that any individual would opt for a social position that is least advantaged, though beneficial. With a qualified first principle, there could be no "least advantaged individual" in society. Thus, the talented artist, mathematician, or poet, though lacking the wealth of the successful entrepreneur, may not wish to exchange his own mode of work and self-actualization for that of the latter. One should pay attention here to the caveat against interpersonal comparisons of utility. Can we prove that the carpenter who experiences much self-actualization in the art of carpentry is less advantaged than the successful but neurotic (to the point of seeking psychiatric care) entrepreneur? It appears, too, that the principle of liberty supported in this discussion might also resolve the question of equality often debated by theorists of rights. I reiterate the point made above that individuals are not in principle resentful of inequalities in income and wealth if it is perceived that opportunities to acquire such are equal.

Thus, I am inclined to believe that the issue of distributive justice entailing transfer of assets, except for the purpose of the funding of public goods, is not as important for the optimal society as some theorists are wont to believe. Accordingly, one would want to acknowledge the irrelevance of discussions on Pareto-optimal social-welfare functions. The idea of equality within society is, therefore, most meaningfully restricted to equality of opportunity to pursue various forms of liberty-enhancing activities. The point is that in contemporary society, whether of liberal democratic or statist structure, individuals in the former are concerned primarily with equal opportunities to invest in their own human capital and the eventual possibility of self-actualizing work; individuals in the latter are concerned principally with self-actualization in the forms of the traditional political freedoms of the former.

Thus, theorists of Marxist persuasion consider the right to employment and other economic rights principal rights due to individuals, while less attention is paid to those rights espoused by those of more liberal outlook, such as their right to free speech, free assembly, and the like. In fact, a key qualification required of those who seek political asylum in the liberal democracies is that the rights they seek to redress in their new societies are political rather than economic in nature. Given these two sets of emphases, I believe that the social collectivity that guarantees both kinds of rights is on the path toward ensuring economic and political optimality. In fact, it would seem that the usual qualitative distinction made between political and economic rights is unjustified since it appears evident that political rights constitute the instrument whereby economic rights are effected. Thus, discussions or controversy over political rights implicitly entail considerations about economic rights. Given the above, the question remains as to the viability of a legally enforced set of rights that guarantees not only employment but all of those rights strongly argued for by the liberal theorist.

The timeliness of this observation is demonstrated by the ongoing experimentation with the principles of liberalism on the part of the governing authorities in the statist societies of the Soviet Union and China. The ideas of *glasnost* and *perestroika* now debated in the former society are no doubt inspired by a consideration of those rights strongly endorsed by the theorist of liberalism. Similar kinds of debates are also current in the centrally controlled society of China. Yet one discerns here a curious phenomenon: the advocate of liberalism usually assumes that there is a qualitative distinction between political rights and economic rights, yet at the same time it is argued that political liberalization in the state-controlled society cannot be truly effected without liberalizations in the economic sphere. At the same time, workers in the liberal democracies have sought to advance the idea of economic rights through the exercise of those political rights supposedly characteristic of the liberal democracies.

It appears, therefore, that the ideas of political and economic rights are closely related. In fact, the reluctance of the neoclassical economist to view the notions of political and economic rights as being contextually related possibly derives from the recognition that a market economy requires less than full employment to serve as a bulwark against a possibly ruinous inflation.

Although I believe that a coordinated effort between government and firms could help implement the idea of the individual's right to employment, I hold also that a necessary condition for so doing is proper deliberation on the question of investments in human capital. In this regard, it would seem that training in a set of specific and general skills during the years of compulsory education would be of much importance.[15] The individual who is trained in a set of skills both for self-employment and otherwise would present to the marketplace an optimal human capital in the sense of being more easily employable and potentially productive.

But those who own capital would prefer, it seems, to deal with a work force for which investments in human capital do not produce workers who can adapt quickly to changes in the economic environment. Quite obviously, the owners of capital would prefer to preserve their monopoly thereon. Apparently, a well-trained adaptive labor force is a potential threat to those who own captial since it opens up the possibility of the equalization of capital ownership. But a society in which human capital investments are optimized is potentially a more productive society. This is a consideration that the neoclassical theorists treat with much discretion.

At this point, it would be instructive to recapitulate what I believe has been established above before a final discussion on a revised theory of the micro- and macroeconomy.

1. I have argued that individual decision making in the area of economics entails the appeal to normative schedules of choice. In this regard, economic decision making is similar to decision making in other normative areas of decision making such as ethics.

2. The question, then, is what considerations should the theorist of economic decision making appeal to in order to establish optimal normative theories of choice both for the individual and for society as a whole. Neoclassical theory does possess a theory of optimal decision making, but (a) it erroneously characterizes it as cognitive in the scientific sense, and (b) it restricts it only to individual behavior—neoclassical welfare economic theory being qualitatively distinct—thereby precluding the formulation of a general theory applicable to all individuals within society.

3. Granted that individuals engage in economic decision making within the context of society, questions on the structure of such societies in terms of the distribution of those freedoms which allow human expression in the form of political and economic decision making necessarily arise.

4. This intellectual concern derives from the fact that human beings are disposed, on account of the structure of their brains, to seek to effect a multiplicity of thoughts. In this sense, it may be appropriate to characterize the human being as "the freedom seeking organism."

5. Neoclassical economic theory is deficient in that it seeks to settle normative questions of choice by appeal to social welfare functions, which are theoretically inoperable because of *a priori* proscriptions against interpersonal comparisons of utility.

6. A theory of optimal economic decision making would include not only propositions prescriptive of economic choice but also a set of assumptions descriptive of contractarian rights for individual agents within society. Given the priority that individuals grant to freedom as the *sine qua non* for human expression, such a theory would seek to establish a contractarian framework in which individual liberties are consensually maximized.

7. A theory of optimal economic decision making would entail, therefore, analyses concerning the structure of the social collectivity in terms of rights-supported freedoms. These would include not only the freedoms of classical liberalism but also an extended set of freedoms concerning the self-actualization of the individual in exchange value generating work.

The above discussion was concerned to lay the groundwork for a theory of optimal decision making for the individual within the social collectivity. It has been shown that such a theory would be optimally operative within the framework of a rights-based society. Such rights would guarantee equilibria of maximal amounts of individual liberties within society. The following discussion would be concerned to explore the structures of a revised micro- and macroeconomic theory.

Revised Microeconomic Models

The following discussion constitutes the formulating of a set of prescriptive criteria necessary for the formulation of microeconomic models of the economy. The practical purpose of these models would be to serve as a policy guide for optimal decision making within the social collectivity. It was stated above that the contours of an optimal social collectivity are defined by a set of legal constraints principally concerned to protect individual rights. In constructing these models, the theorist would appeal to the most appropriate mathematical techniques adapted from research areas such as operations research, industrial engineering, and the like. More specifically, the theorist would appeal to mathematical techniques to construct decision-making models founded on relevant empirical data such as revenues, costs, capital structures, and so on. The following lists a set of recommended operational procedures that one might consider in formulating theories of decision making for the microeconomy.

1. I begin with dispensing altogether with the postulate of rationality and the concept of utility, two concepts that have been the source of much difficulty for theorists of economics. Recall the historical rejection of cardinal utility in favor of an equally problematic ordinal utility, also Simon's more recent attempts to replace substantive rationality (i.e., classical rationality) with bounded and procedural rationality. The concepts of rationality and utility are to be considered inoperable in this model since they do not serve any heuristic function. The alternative theory consists of sets of optimizing models that determine the most efficient choice-paths given certain prescribed ends.

These models would be structured on appropriate aspects of linear and nonlinear programming, such as goal programming for multiple objective functions, multicriteria decision making, and quadratic programming. One recognizes that in place of talk about utility or profit maximization, one must effect a paradigm shift and talk instead of maximizing and minimizing objective functions subject to sets of constraints.

The normative or policy-oriented aspect of this approach is summed up by the following criteria for microeconomic decision making:

i. The theory of decision making entails the experimental formulation of strategies for the attainment of certain preset goals within the context of relevant constraints.
ii. Given that an infinite set of strategies could be devised for the attainment of goals, the theory of decision making is concerned to establish criteria whereby those strategies considered optimal could be identified.

2. The test of the theory would entail the formulation of case studies of the empirical behavior of firms and individual agents within the general economy. Firms that have demonstrated consistently optimizing behavior should be analyzed and models of their operating procedures constructed. It should be noted that "optimizing behavior" need not mean maximization of profits; it means only "achieving of specific goals subject to desired costs given recognized constraints." Of much importance, too, would be diagnostic analyses of the operational procedures of those firms which had not been able to optimize over time. The goals of empirical analysis in this context are to establish close diachronic links between optimizing theories and their operational implementations in the general economy.

Models of individual agent choice would be constructed along lines similar to those of the firm. For this purpose, it would be heuristically useful to regard each individual agent as a production complex whose structure is determined by sets of cost-bearing inputs and goal attainment outputs. The cost-bearing inputs include investments in human capital, consumption expenditures, and the like, and the goal-attaining outputs would be determined by items such as the exchange value received by the agent, behavioral evidence of psychic satisfaction, and so on.

The choice paths of individual agents would then be evaluated by appeal to theories of multicriteria decision making and goal programming. Optimal decision making at the individual level would entail taking into consideration real constraints, costs, and the possibility of attaining desired goals. The simplistic notion of the maximization of utility subject to budget constraints obviously cannot do justice to the particular notions of optimality harbored by each individual agent. For example, the optimal goal of a given individual over a given time path could be that of a frugal existence in a rural environment, while that of another might be the accumulation of enormous material wealth. Thus, this normative theory of agent decision making cannot be tested as if it represented the behavioral choices of all individuals—it can serve only as a prescriptive guide for individual agents given that optimality in terms of goals and the paths for attaining those goals can vary so greatly.

In sum, the theorist of microeconomic decision making could be viewed in much the same way as the physician who is trained to regard the attainment of good health as an optimal goal and whose social function is to prescribe to individuals who seek optimal health the means of attaining it within particular constraints. But here again, the specified criteria of what should constitute optimal health would vary from individual to individual.

A Revised Macroeconomic Theory

Macroeconomic theory in general is concerned to analyze the total economic activity of society not only in terms of the simultaneous functioning of its several microeconomic units, but also in terms of governmental activity. The role of the latter in general macroeconomic activity is somewhat distinct from that of the microeconomic sector of the economy in that government not only engages in productive activity, but also seeks to regulate the behavior of the microeconomic units by means of fiscal or monetary policies.

Perhaps one of the major debates in contemporary economic theory is that concerning the role of government in the general economy. The Keynesian school of thought, dominant over the last three decades in the industrialized societies, has been challenged by those who believe that the role of government should be minimal in the functioning of the economy. Thus, libertarians argue that government intervention in the economy should be concerned mainly with the protection of the individual against force, fraud, and possible breaches of contracts.[16] Public choice theorists, as seen, also argue strongly against governmental intervention in the economy for purposes of revenue or capital redistribution since those measures not only militate against the idea of free enterprise but also lead to market inefficiencies.[17] A corollary to these new arguments is that an increase in the economic power of the government necessarily leads to an increase in its political power, hence the ensuing risk of the diminution of liberties for the governed.

On the other hand, those theorists in favor of an expanded role of government in the economy not only for purposes of monitoring the economy but also for the redistributing of income believe that the economic activity of the government cannot be divorced from questions about social justice and equity.[18] Those theorists who believe that questions of equity should be accorded priority in macroeconomic policy believe that the optimal policy is one in which great differences in individual wealth are reduced by means of government policy. The mixed economies of the Scandinavian countries are empirical examples of such. Government-sanctioned rights are mandated not only for individual investment in human capital for the preadult years, but also for access to medical services and housing. It is evident that such societies attempt to establish some socially acceptable equilibrium of economic and political rights.

The macroeconomic theory I propose in the light of the above discussion is one that is grounded in the ideas of the social collectivity and individual rights. The basis for this was analyzed above. An optimal macroeconomy, therefore, is one in which the government is delegated the role of monitor of the economy so that individual economic and political rights be maintained.

More specifically, the monetary policy of the government in the macroeconomy would be one of promoting even economic growth and development. The general banking system in the economy would include not only privately owned banks, but also government banks whose function would be to ensure public accessibility to long-term loans at low interest specifically for purposes of investment in long-term durable goods such as human capital, housing, and so on.

The fiscal policy of the government would be concerned with setting the conditions for the implementation of the longer-term goals such as the expenditure of tax revenues on social infrastructures, universal investment in human capital, and the establishing of productive enterprises that would employ those individuals who had lost employment in enterprises experiencing structural changes. The basis for this, as argued, is that any social collectivity that takes individual rights seriously would include the right to productive labor as a primary right. Thus, government investment in training programs, the relocation of surplus labor from one sector of the economy to other sectors where the demand for labor is growing, and new investments in targeted areas would do much to ensure that the right to productive labor be universally recognized. It should be recognized that the implementation of such policies would be the responsibility of special government agencies that would appeal to techniques of mathematical goal programming and industrial engineering.

An optimal fiscal policy would also ensure that tax revenues be utilized not only for social programs such as pensions, health services, and so on, but also to facilitate the individual's access to capital through stock purchases and the like. Individuals would be obliged not only to acquire capital shares in their own enterprises, but elsewhere according to their preference. The point here is that an optimal fiscal policy requires not only that the individual exercise his or her right to productive work, but also the right to his or her own economic surplus. Recall that it is the profits—in reality the net surplus—of the enterprise that are necessary for the growth of a firm's capital stock. And one can settle the question of the distribution of a firm's surplus by determining it as a function of wages. Of course, there is a wide variety of ways in which the individual employed by a particular enterprise could share in society's capital stock. For example, it could be legally mandatory that the majority of the stock shares of any enterprise be owned by its workers. This would no doubt serve as incentive for increased productivity on their part. Other experiments in this regard are possible.

The role that general macroeconomic fiscal policy could play in the productive individual's right to his or her own surplus is that the distribution of this surplus would be supervised by the government in much the same way that tax liabilities are enforced. It is seen, therefore, how an optimal macroeconomic fiscal policy founded on the enforceable right of productive labor for each individual aims at three goals: (1) adequate government investment in human capital and other socially necessary infrastructures, (2) full employment, and (3) the increase and distribution of capital stock throughout the economy.

This theory contrasts, of course, with that of the orthodox neoclassical school, which regards less than full employment as optimal—a safeguard against a much-feared inflation. The basic reason for this stance is that neoclassical theory tacitly accepts the thesis that the entrepreneur minimizes labor costs most when the supply of labor is most competitive—in other words, when the supply of labor is greater than the demand for it. This is understandable since the orthodox model approaches the macroeconomic problem mainly from the point of view of investment and financial institutions rather than from that of full employment, capital distribution, and balanced growth. Note, too, that the orthodox microeconomic theory emphasizes the maximization of profit as the goal of entrepreneurship.

I suspect that the macroeconomic theory advocated in this discussion would be stoutly resisted by those theorists who, under the guise of scientific objectivity, would prefer to maintain the status quo with its great disproportionalities in capital distribution throughout the societies of mixed economy. But balanced growth and social stability are to be least expected from those societies in which full employment is discouraged and capital ownership is monopolized by a minority. It should be noted parenthetically that the so-called Keynesian revolution was really not an attempt to create the conditions for an optimal economy but to establish the intellectual basis for the rescue by the government of a potentially anarchistic economic system. Recall that Keynes's main premise was the politically important observation that labor costs were no longer susceptible to downward pressures because of the growing power of the worker trade unions.

Similar criticisms apply to those societies in which full employment is enforced by the law, but workers themselves are unable to own capital in any meaningful way. In those societies, the bulk of productive capital is regarded as "state property," managed but not owned by the directors of state enterprises. Once again, there is a situation in which workers' surplus is confiscated not by the owner of capital, but by the state. This tenuous situation is maintained by the general restriction on political rights of individuals enforced by the state. It seems to me that when the state becomes the sole entrepreneur, the right to employment is guaranteed but not the individual right to own capital and the right to portions of one's surplus as worker. In this context, the political freedoms of individuals are severely restricted.

But on the other hand, the so-called invisible-hand approach advocated by neoclassical theory, as evidence shows, leads to anarchy, and hence the need for the guiding role of the government. The important epistemological question at this point is whether the market system guided by the state, not for the maximization of profits, but for the optimization of welfare, is possible. I have argued above that this is indeed possible so long as the structure of the social collectivity is defined in terms of enforceable economic and political rights for all its members. I believe that societies such as Iceland, Japan, and Norway offer glimpses of this possibility.

Some will argue that the impact of this analysis is negative in the sense that it reduces economics to the level of politics, an activity burdened by ideological dispute. But the question of rights involves normative considerations, and quite remarkably there is a growing consensus as to the nature of such rights. And a key point in this text is that there are major grounds for arguing that economic rights constitute an important aspect of the general concept of rights.

In fact, it is this point concerning the problematic of political and economic rights that has recently become the issue in the formulating of the ideas of *glasnost* and *perestroika* in the Soviet Union, for example. These two concepts embody not the reexamination of the idea of political rights but also that of economic rights in the communist world. Consider, too, the fact that although there is universal agreement on the need for increased political rights in the communist world, there is much debate concerning individual economic rights. There has been agreement that the idea of the state control of the means of production must be reevaluated. But the key question then is what should become of state property? Some argue for "privatization" in the Western sense of that term, but are immediately reminded of the painful costs of unemployment and social dislocation that this would produce.

It is at this juncture, I believe, that the intriguing question about worker ownership of the means of production becomes relevant. Maximal political freedom can be supported only by maximal economic freedoms. It is only then that the state would begin to wither away. It is more than just another historical curiosity to wonder whether the proper implementation of such (maximal, political, and economic freedoms) would engender other specters. Note in closing that all of the ideas discussed above, as I have argued, could be best experimented on in a context rid of the notion of economics as a positive science.

A fitting conclusion to the analysis in this chapter is that, contrary to its avowed methodology, the neoclassical research program is minimally concerned with empirical analysis and maximally concerned with refinements of the formal models of individual agent theory and general equilibrium analysis. The basic assumptions of neoclassical microeconomic theory are constituents of a normative framework populated only by "rational" agents. This peculiar relationship to the empirical world is determined by the postulate of rationality and the *ceteris paribus* proviso. In the case of a genuine scientific theory, basic axiomatic assumptions are linked deductively to the empirical world by means of bridge principles. It is in this context that one speaks of the predictive test implications of a theory as a confirmatory warrant of its theoretical assumptions. This is not the case when the test implications of a theory are not deductively related to its assumptions—as is the case with neoclassical economic theory.

Conclusion

The bulk of this text was taken up with examining the claims of neoclassical economic theory to scientific status. Given contemporary views on the nature of scientific theory, I examined neoclassical economic theory in terms of both its historical and contemporary phases. I demonstrated that the cardinal theory of utility that formed the foundation for early neoclassical theory foundered on account of its inability to measure utility in any acceptable scientific way. Its substitute, the ordinal theory of utility, was shown to be equally unacceptable. The scientific pretensions of ordinal utility theory and its correlate, revealed preference theory, were shown compromised by the normative structure of the foundational postulate of rationality. The unscientific nature of ordinal utility theory was further shown to be reinforced by the insulating role played by the *ceteris paribus* proviso.

This general critique was extended not only to the neoclassical theory of individual agent choice but also to general equilibrium theory and positive neoclassical welfare economic theory. Given the general dissatisfaction with neoclassical theory, a number of alternative theories have been proposed, but the problem with the latter is that, with few exceptions, they are founded on the premise that an objective science of economics is still possible despite its present failings. I pointed out the shortcomings of those theories and argued that on account of the nature of human decision making, no analysis of it could be scientific in the way in which the natural sciences are scientific. Mental states that must be invoked to explain behavior are just not subject to empirical analysis. The attempts by theorists to establish explanatory theories by appeal to heuristic concepts such as rationality were shown to be unsuccessful. The point is that "rationality" plays a normative role similar to that of "goodness" in ethical theory.

The sociologist can indeed record the behavior of individuals in terms of cultural norms of "goodness," "badness," "deviancy," and so on, but he or she must recognize that theories of behavior founded on such concepts are necessarily normative. Similarly, the neoclassical theorist who embraces a particular notion of rationality and grounds his or her theories on such a notion is certainly formulating a normative theory. My analysis showed that the neoclassical theorist of economic behavior is confronted with the dilemma of restricting his or her analysis to a case-by-case taxonomy of individual agent choice, given the inaccessibility to mental states, or grounding his or her explanatory theories on the normative heuristic of rational choice. Neither alternative yields scientific results.

In sum, this analysis demonstrated that no theory of economic decision making could fail to be normative. Given this, the question then was what normative theory theorists of economic behavior should seek to formulate. I attempted to show that an optimal theory of economic decision making was one in which individuals were guaranteed rights to maximal liberties in both the political and economic spheres of society. It was then argued that it was the primary right to productive self-actualizing work that provided the means to the realization of that stock of political liberties necessary for optimal decision making within society. There is yet hope for the "dismal science" in its continuing efforts to articulate the conditions for human well-being.

Notes

Chapter 1. Introduction

1. One might consider the following works as representative of the growing number of critical evaluations of contemporary neoclassical theory. Mark Blaug, *The Methodology of Economics* (Cambridge: Cambridge University Press, 1980); Homa Katouzian, *Ideology and Method in Economics* (New York: New York University Press, 1980); Daniel Bell and Irving Kristol, eds. *The Crisis in Economic Theory* (New York: Basic Books, 1981); S. J. Latsis, ed., *Method and Appraisal in Economics* (Cambridge: Cambridge University Press, 1976); D. N. McCloskey, *The Rhetoric of Economics* (Madison, Wisconsin: The University of Wisconsin Press, 1985); and Patrick J. O'Sullivan, *Economic Methodology and Freedom to Choose* (London: Allen and Unwin, 1987).

2. It should be noted that the scientific study of animal behavior (nonhuman) has been relatively successful only because much of animal behavior is determined by instinctual drives usually expressed in the form of predictive responses to particular stimuli, hence making it easier for the researcher to explain behavior purely in terms of what has been observed. Clearly, nonhuman animals do not display the wide variety and plasticity of subjectively conscious decision making as humans do. It is for this reason that models of ethical behavior and rationality are the human substitutes for drives and instincts in animals.

3. Appropriate examples of ongoing research on expected utility theory and its attendant rules of rational choice are as follows: Paul J. H. Schoemaker, "The Expected Utility Model: Its Variants, Purposes, Evidences and Limitations," *Journal of Economic Literature* 20 (June 1982): 529–63; David Grether and Charles Plott, "Economic Theory of Choice and the Preference Reversal Phenomenon," *American Economic Review* 69, no. 4 (1979): 623–38; Amos Tversky and Daniel Kahneman, "The Framing of Decisions and the Psychology of Choice," *Science* 211 (1981); M. Machina, " 'Expected Utility' Analysis without the Independence Axiom," *Econometrica* 50 (1982): 277–323; and Charles F. Manski, "Ordinal Utility Models of Decision Making under Uncertainty," *Theory and Decision* 25, no. 1 (1988): 79–104.

4. See Amartya Sen, *On Ethics and Economics* (Oxford: Basil Blackwell, 1987), and Alan Hamlin, *Ethics, Economics and the State* (New York: St. Martin's Press, 1986), as examples of the growing intellectual interest in the relationship between economic decision making and ethics.

5. See E. J. Mishan, *Introduction to Normative Economics* (New York: Oxford University Press, 1981), p. xvii. Mishan writes:

> An economics that was entirely positive might be regarded as a prestigious intellectual pursuit. But it would not command the respect of the general public. If therefore one observes that economics appears today to be held in high esteem—rather more esteem, I

should say, than is warranted by its record—it is simply because economics is, after all, believed also to be a normative study, one that informs social policy. For better or worse, a prescriptive economics is in active service in all civilized societies. Economic growth, high employment, and price stability are common objectives common to all governments.

Chapter 2. Knowledge and the Theory of Science

1. See Robert McRae, "The Unity of the Sciences: Bacon, Descartes, Leibniz," in *Roots of Scientific Thought*, ed. Philip Weiner and Aaron Noland (New York: Basic Books, 1957), p. 390.
2. Isaac Newton, *Principia Mathematica*, ed. Florian Cajori (Berkeley: University of California Press, 1960), p. 547.
3. Larry Laudan, *Science and Hypothesis* (Boston: D. Reidel Publishing Company, 1981), pp. 9–15.
4. Carl G. Hempel and Paul Oppenheim, "Studies in the Logic of Explanation," *Aspects of Scientific Explanation* (New York: The Free Press, 1965).
5. Sylvain Bromberger, "Why Questions," in *Readings in the Philosophy of Science*, ed. Baruch Brody (Englewood Cliffs, N.J.: Prentice Hall, 1970), pp. 66–87.
6. Michael Scriven, "The Temporal Asymmetry of Explanations and Predictions," in *Philosophy of Science*, The Delaware Seminar, 2 vols., ed. B. Baumrin (New York: Wiley, 1962–63), pp. 97–105.
7. Wesley Salmon, *Scientific Explanation and the Causal Structure of the World* (Princeton: Princeton University Press, 1984), pp. 48–55.
8. See Paul Feyerabend's *Against Method* (London: New Left Books Press, 1975).
9. See Larry Laudan, *Progress and Its Problems* (Berkeley: University of California Press, 1981), for an instrumentalist approach to the history of science. Laudan's error, it seems to me, is to argue that instrumentalism or mere problem solving is a sufficient condition for some research enterprise to be regarded as a science.
10. Feyerabend, *Against Method*.
11. See, for example, E. Roy Weintraub, *General Equilibrium Analysis* (Cambridge: Cambridge University Press, 1985), pp. 27–33.
12. Karl Popper, *Popper Selections*, ed. David Miller (Princeton: Princeton University Press, 1985), pp. 101–17.
13. Ibid., p. 128.
14. Ibid., pp. 128–29.
15. Blaug, *The Methodology of Economics*, p. 260.
16. See D. Wade Hands, "Karl Popper and Economic Methodology," *Economics and Philosophy* 1 (1985). Hands argues in this paper that Popper himself may have had in mind only the natural sciences when he formulated his falsificationist philosophy. He claims that in his writings on economics Popper chose instead to appeal to "situational logic," or "situational analysis." According to Hands, "the fundamental tenet of the program is that we should and must seek explanations of social behavior in terms of the 'situation' in which the agents find themselves" (p. 86). Hands's aim here, no doubt, is to shield neoclassical economic theory from the popular falsificationist Popperian methodology with the thesis that the latter applied only to the methodology of the natural sciences. Note, too, Marc

Blaug's response to Hands in "Comment on D. Wade Hands, Karl Popper and Economic Methodology: A New Look," *Economics and Philosophy* 1 (1985): 286–88. He appears to agree with Hands that falsificationism may not in fact be the methodological tool most economists appeal to in their appraisal of economic theory.

17. S. J. Latsis, "A Research Programme in Economics," in *Method and Appraisal in Economics*, ed. Spiro J. Latsis (Cambridge: Cambridge University Press, 1976), p. 22.

18. Ibid.

19. A. Leijonhufvud, "Schools, Revolutions and Research Programmes," in *Method and Appraisal in Economics*, ed. Latsis, pp. 65–108; Weintraub, *General Equilibrium Analysis*.

20. Laudan, *Progress and Its Problems*.

21. T. W. Hutchison, "History and Philosophy of Science and Economics," in *Method and Appraisal in Economics*, ed. Latsis. T. W. Hutchison's observations here serve as a useful contrast to the optimistic theorizing of those theorists who believe that with a few adjustments here and there economics could be transformed into a respectable scientific discipline. He writes:

> But in spite of a tendency to some kinds of improvement, new and old forms of the kind of intellectual malpractices which Popper's prescriptions were designed to combat are still widespread in economics: verbalism, conventionalist, and immunizing stratagems, and the erosion of testable formulations and testing. It has been taken for granted by many economists and their methodologist RPOs, that no criticisms of the assumptions underlying models, which would require their testing or testability, need to be needed. (p. 202)

Chapter 3. The Structure and Proof of Scientific Theories

1. Hempel's paper, "The Theoretician's Dilemma," offers a useful discussion of the observational-theoretical problem. See Carl Hempel, "The Theoretician's Dilemma," in *Aspects of Scientific Explanation*, pp. 173–226.

2. Carl Hempel, "Typological Methods in the Natural and Social Sciences," in *Aspects of Scientific Explanation* (New York: The Free Press, 1965), p. 170.

3. Hempel, "Studies in the Logic of Explanation," in ibid., p. 257.

4. Nancy Cartwright, *How the Laws of Physics Lie* (Oxford: Oxford University Press, 1983).

5. John Stuart Mill, *On the Logic of the Moral Sciences* (New York: The Bobbs-Merrill Company, Inc., 1965), pp. 26–27.

6. Auguste Comte, *Philosophie Positive* (Paris: Bachelier, 1830), 1, pp. 15–16.

7. Mill, *On the Logic of the Moral Sciences*, pp. 320–21.

8. Ibid., pp. 26–27.

9. Comte, *Philosophie Positive*, pp. 16–17.

10. See Max Weber, "The Interpretive Understanding of Social Action," in *Readings in the Philosophy of the Social Sciences*, ed. May Brodbeck (New York: The Macmillan Co., 1968), p. 33.

11. Ibid., p. 23.

12. Karl Popper, *The Poverty of Historicism* (New York: Harper Torchbooks, 1964), pp. 130–31.

13. Ernest Nagel, "The Value-Oriented Bias of Social Inquiry," in *Readings in the Philosophy of the Social Sciences*, ed. Brodbeck, p. 99.
14. Ibid., p. 43.
15. Carl Hempel, "Typological Methods in the Social Sciences," in *Philosophy of the Social Sciences*, ed. Maurice Natanson (New York: Random House, 1963), p. 219.
16. Ibid.
17. Gunnar Myrdal, *Objectivity in Social Research* (London: Gerald Duckworth, 1970).
18. Thomas Kuhn, *The Structure of Scientific Revelations*, 2d ed. (Chicago: The University of Chicago Press, 1972), pp. 160–61.
19. See Gerald Doppelt, "The Philosophical Requirements for an Adequate Conception of Scientific Rationality," *Philosophy of Science* 55 (1988): 104–33, and Frans Gregersen and Simo Køppe, "Against Epistemological Relativism," *Studies in History and Philosophy of Science* 19, no. 4 (1988): 447–87. The latter comment on the influence of the idea of epistemological relativity in the formulation of theories of scientific change in history, yet recognize that if the scientific enterprise is to be explained as progressive in the cumulative sense, there must be some discernible thread of rationality connecting its successor and precursor theories.
20. Kenneth J. Gergen's highly illuminating article, "Correspondence versus Autonomy in the Language of Understanding Human Action," points out from the standpoint of the psychologist the general epistemological difficulties encountered by the theorist in establishing even an adequate theoretical framework for purposes of explaining human behavior. See Gergen's article in *Metatheory in Social Science*, ed. Donald Fiske and Richard Schweder (Chicago: University of Chicago Press, 1986), pp. 136–62. In this regard, the appeal by some theorists of economics to behavioral psychology as a means of anchoring in a quantitative sense the theory of decision making would be unrewarding.
21. See, for example, the recent papers by Laudan, Rosenberg, Doppelt *et al.*, on the issue of normative naturalism in *Philosophy of Science* 57 (1990). See, too, the same topic being discussed by Laudan and Siegel in *Studies in the History and Philosophy of Science*, Vol. 21, No. 2, 1990. No working scientist would deny that there are normative rules of procedure in the testing of hypotheses, while recognizing that such rules are not arbitrary. They are chosen over alternatives simply because their application yields results that help determine whether experimental anticipations are justified in terms of actual observable occurrences.
22. See Karin Knorr-Cetina's, *The Manufacture of Knowledge* (New York: Pergamon Press, 1981) for an involved study of the experimental methods employed in scientific research.

Chapter 4. The Classical and Neoclassical Methodology

1. Jeremy Bentham, *Jeremy Bentham's Economic Writings*, ed. Werner Stark (London: Allen and Unwin, Ltd., 1952), 1: 102–7.
2. Ibid., 3: 318.
3. Ibid., 1: 103–19.
4. See James Mill, *Selected Economic Writings*, ed. Donald Winch (Chicago: University of Chicago Press, 1966).

5. Some theoreticians of economics would take issue with the thesis that Newtonianism was the dominant exemplar in the formulation of neoclassical theory. Philip Miroski, for example, writes that "recourse to the history of mathematics and physics shows that the characterization of neoclassical economics as 'Newtonian' is both inept and misleading." See Philip Mirowski, "Physics and the Marginalist Revolution," *Cambridge Journal of Economics* 8 (1984): 365. Mirowski would argue instead that "the adoption of the 'energetics' metaphor and framework of mid-nineteenth century physics is the birthmark of neoclassical economics, the Ariadne's thread which ties the protagonists, and which can lead us to the fundamental meaning of the neoclassical research programme" (p. 365). I am inclined to believe, though, that Mirowski's claim cannot be truly substantiated since it is indeed possible to view the energetics research school as a subparadigm in the major dominant Newtonian paradigm of the time. It was merely a continuation and exploration of the basic principles of Newtonianism as exemplified in Newton's classical laws. Mirowski would appear to base his claim on an evident incompatibility between energetics theory and Newtonianism. This is not the case; rather, energetics theory is entailed by Newtonian theory.

6. Vilfredo Pareto, *Manual of Political Economy* (New York: Augustus M. Kelley, 1971), p. 10.

7. W. S. Jevons, *The Theory of Political Economy* (New York: Augustus M. Kelley, 1965), pp. vi–vii.

8. Ibid., pp. 59–60.

9. Leon Walras, *Elements of Pure Economics*, trans. William Jaffe (Homewood, Illinois: Irwin Inc., 1954), p. 71.

10. Ibid., p. 47.

11. Ibid., p. 185.

12. Alfred Marshall, *Principles of Economics* (New York: Macmillan, 1961), p. 35.

13. Ibid., p. 43.

14. J. von Neumann and O. Morgenstern, *Theory of Games and Economic Behavior* (New York: John Wiley and Sons, Inc., 1967).

15. Ibid., p. 18.

16. Ibid.

17. See, for example, William Baumol, "The Neumann-Morgenstern Utility Index—An Ordinalist View," *Journal of Political Economy* 59 (February 1951): 61–66.

18. von Neumann and Morgenstern, *Theory of Games and Economic Behavior*, p. 29.

19. See Paul J. H. Schoemaker, "The Expected Utility Model: Its Variants, Purposes, Evidence and Limitations," *Journal of Economic Literature* 20 (1982): 529–63, for a useful discussion on the foundational influence of the von Neumann–Morgenstern utility theory on the theory of expected utility. In attempting to demonstrate the relationship between modern ordinal expected utility theory and the von Neumann-Morgenstern theory, Schoemaker writes: "from a measurement perspective, NM utility theory is cardinal in that its utility scale has interval properties. However, from a preference perspective, NM utility theory is ordinal in that it provides no more than ordinal rankings of lotteries" (p. 533).

20. Lionel Robbins, *The Nature and Significance of Economic Science* (London: Macmillan and Co., Ltd., 1932), p. 138.

21. P. A. Samuelson, "The Empirical Implications of Utility Analysis," in *The Collected Scientific Papers of Paul A. Samuelson*, vol. 1, ed. Joseph Stiglitz (Cambridge: The M.I.T. Press, 1966), p. 21.

22. J. R. Hicks, *Value and Capital* (Oxford University Press, 1974), p. 19.
23. Ibid., p. 20.
24. Ibid.
25. P. A. Samuelson, "A Note on the Pure Theory of Consumer Behavior," in *The Collected Scientific Papers of Paul A. Samuelson*, 1: 3-4.
26. Ibid., p. 4.
27. Stanley Wong, *The Foundations of Paul Samuelson's Revealed Preference Theory* (London: Routledge and Kegan Paul, 1978).
28. T. W. Hutchison, *The Significance and Basic Postulates of Economic Theory* (London: Allen and Unwin, 1938), chapters 2, 3, and 4 especially.
29. F. Machlup, "The Problem of Verification in Economics," in *The Methodology of Economics and Other Social Science* (New York: New York University Press, 1967), pp. 143-47.
30. T. W. Hutchison, "Professor Machlup on Verification in Economics," *Southern Economic Journal* 22 (April 1956): 476-83; F. Machlup, "Rejoinder to a Reluctant Ultra-Empiricist," *Southern Economic Journal* 22 (April 1956): 483-93. Note that this paper is also published as "Terence Hutchison's Reluctant Ultra-Empiricism," in Machlup, ed., *The Methodology of Economics*, pp. 493-503.
31. Ernest Nagel, "The Subjective Nature of Social Subject Matter," in *Readings in the Philosophy of the Social Sciences*, ed. Brodbeck, p. 43.
32. Henry Margenau, "What Is a Theory," in *The Structure of Economic Science*, ed. Sherman Krupp (Englewood Cliffs, N.J.: Prentice Hall, 1966), p. 38.

Chapter 5. Ordinal Utility Theory and Contemporary Neoclassical Economics

1. Apparently the ordinalist program of the neoclassicals has been able to establish itself as the dominant scientific research paradigm in contemporary economics. Convinced of the scientific nature of their methodology, neoclassical economists speak unreservedly of positive economics and normative economics as distinct branches of economic theory. Consider the following standard observation by a prominent neoclassical theorist.

> In so far as it is a positive, that is, explanatory science, economics must analyze the behavior of agents who enjoy some freedom but are subject to the constraints imposed on them by nature institutions. In so far as it is a normative science, economics must also investigate the best way of organizing production, distribution and consumption. It must give the conceptual tools which enable us to assess the comparative advantage of different forms of organization. In its pursuit of this double activity, positive and negative, our science has come to attribute a central role to the prices which regulate the exchange of goods among agents. (E. Malinvaud, *Lectures on Microeconomic Theory* [London: North Holland Publishing Co., 1972], p. 2.)

2. See Vivian Walsh, *Introduction to Contemporary Economics* (New York: McGraw Hill, 1970), as an example of an almost purely choice-theoretic approach to individual decision making.
3. See James Henderson and Richard Quandt, *Microeconomic Theory* (New York: McGraw-Hill, 1971), p. 45.
4. Blaug, *The Methodology of Economics*, p. 167.

5. Stanley Wong, *The Foundations of Paul Samuelson's Revealed Preference Theory* (London: Routledge and Kegan Paul, 1978).
6. Ibid., p. 121.
7. For example, Wayne J. Shafer has shown that the transitivity assumption is not necessary to establish a consistent theory of demand. See Wayne J. Shafer, "The Nontransitive Consumer," *Econometrica* 42, no. 5 (1974): 913–19. Yet incidentally, Robert Russell, a consultant for Shafer's article, argues that without the assumptions of asymmetry and transitivity, "The consumer's preferences would go around in circles, reflecting a type of inconsistency which would make the problem of predicting the consumer's behavior extremely problematical." See Robert Russell and Maurice Wilkinson, *Microeconomics: A Synthesis of Modern and Neoclassical Theory* (New York: Wiley and Sons, 1979), p. 14.
8. See Blaug, *The Methodology of Economics*, pp. 159–73, for a critical analysis of demand theory in terms of its empirical content.
9. Ibid., p. 169.
10. See Russell and Wilkinson, *Microeconomics*, p. 49.
11. Ibid., p. 70.
12. Blaug, *The Methodology of Economics*, p. 165.
13. M. Friedman, "Methodology of Positive Economics," in *Essays in Positive Economics* (Chicago: University of Chicago, 1953), pp. 3–43.
14. P. Samuelson, "Problems of Methodology—A Discussion," *American Economic Review (Proceedings)* 53 (May 1963): 231–36.
15. Friedman, "Methodology of Positive Economics," pp. 8–9.
16. T. Koopmans, *Three Essays on the State of Economic Science* (New York: McGraw Hill, 1957).
17. E. Rotwein, "On 'The Methodology of Positive Economics,'" *Quarterly Journal of Economics* 73 (1959): 554–75.
18. H. Simon, "Problems of Methodology—A Discussion," *American Economic Review (Proceedings)* 53 (1963): 231–36.
19. P. Samuelson, "Problems of Methodology—A Discussion."
20. Ibid., pp. 233–34.
21. P. Samuelson, "Theory and Realism: A Reply," *American Economic Review* 54 (1964): 738.
22. Ibid.
23. Friedman, "Methodology of Positive Economics," p. 16.
24. Ibid., p. 21.
25. See, for example, Hempel's discussion on the concept of rationality in his *Aspects of Scientific Explanation*, pp. 463–87.
26. The point made here is aptly recognized by theorists of the scientific enterprise who seek to establish logical links between directly observable entities (described by observation terms) and indirectly observable entities (described by theoretical terms). The aim here is to demonstrate that theoretical terms do possess empirical content, despite their naming of entities far removed from content, despite their naming of entities far removed from direct observation. See, for example, Hempel, *Aspects of Scientific Explanation*, pp. 173–226.
27. Ernest Nagel, "Assumptions in Economic Theory," *American Economic Association Papers and Proceedings* 53 (1963): 211–19.
28. Ibid., p. 215.
29. Ibid., p. 219.
30. A. Coddington, "Positive Economics," *Canadian Journal of Economics* 5 (1972): 1–15.
31. See Lawrence Boland, "A Critique of Friedman's Critics," *Journal of*

Economic Literature, June 1979, pp. 503-22; W. Frazer and L. Boland, "An Essay on the Foundations of Friedman's Methodology," *American Economic Review* 73 (March 1983): 129-44; and Lawrence Boland, *The Foundations of Economic Method* (London: George Allen and Unwin, 1982).

32. Fritz Machlup, "Paul Samuelson on Theory and Realism," in *Methodology of Economics and Other Social Sciences*, ed. Machlup (New York: Academic Press, 1978), p. 481.

33. Ibid.

34. Ibid., pp. 482-83.

35. Ibid., p. 735.

36. Paul Samuelson, "Professor Samuelson on Theory and Realism: Reply," in *The Collected Scientific Papers of Paul Samuelson*, vol. 3, ed. Robert Merton (Cambridge: M.I.T. Press, 1972), p. 772.

37. Paul Samuelson, "Theory and Realism: A Reply," in Merton, ed., p. 762.

38. Samuelson, "Professor Samuelson on Theory and Realism," p. 772.

39. Stanley Wong, "The F-Twist and the Methodology of Paul Samuelson," *American Economic Review* 63 (1973): 319.

40. Lawrence Boland, "On the Futility of Criticizing the Neoclassical Maximization Hypothesis," *American Economic Review* 71, no. 5 (December 1981): 1036.

41. Ibid., p. 1034.

42. Ibid.

43. Ibid.

44. Ibid.

45. Ibid. Consider, too, Philippe Mongin's useful thesis that "all-and-some" statements could be shown refutable with appropriate translation rules. See Philippe Mongin, "Are All-and-Some Statements Falsifiable after All," *Economics and Philosophy* 2 (1986): 185-95.

46. See A. Y. C. Koo, "An Empirical Test of Revealed Preference Theory," *Econometrica* 30 (1962): 646-64, and A. Y. C. Koo and G. Hasenkamp, "Structure of Revealed Preference: Some Preliminary Evidence," *Journal of Political Economy* 80 (1982): 724-44. See, too, the research results of Amos Tversky and Daniel Kahneman in the following: "Prospect Theory: An Analysis of Decision Under Risk," *Econometrica* 47 (1979): 263-91; "The Framing of Decisions and the Psychology of Choice," *Science* 211 (1981): 453-58. Also, Frederic Schick, "Rationality," *Economics and Philosophy* 3 (1987): 49-66.

47. Boland, "On the Futility of Criticizing the Neoclassical Maximization Hypothesis," p. 1034.

48. See Paul Schoemaker, "The Expected Utility Model," *Journal of Economic Literature* 20 (1982): 531.

49. Blaug, *The Methodology of Economics*, p. 260.

50. See T. W. Hutchison, "History and Philosophy of Science and Economics."

51. See *Method and Appraisal in Economics*, ed. S. Latsis, for a collection of papers that explore the possibilities of interpreting economic theory from the standpoint of Lakatos's methodology of scientific research. See also Weintraub's *General Equilibrium Analysis: Studies in Appraisal*.

52. Paul Feyerabend, "Consolations for the Specialist," in *Criticism and the Growth of Knowledge*, ed. I. Lakatos and A. Musgrave (Cambridge: Cambridge University Press, 1970), pp. 229.

53. Alexander Rosenberg, "Lakatosian Consolations for Economics," *Economics and Philosophy* 2 (1986): 127-39.

54. See Alexander Rosenberg, "A Skeptical History of Microeconomic Theory,"

Theory and Decision 12 (1980): 79–93; "If Economics Isn't Science, What Is It?" *Philosophical Forum* 14 (1983): pp. 296–314.

55. Donald McCloskey, *The Rhetoric of Economic* (Madison: The University of Wisconsin Press, 1985), p. xix.

Chapter 6. General Equilibrium Theory—An Analysis

1. See Leon Walras, *Elements of Pure Economics*, trans. William Jaffe (Homewood, Illinois: Irwin, Inc., 1954); Abraham Wald, "On Some Systems of Equations of Mathematical Economics," *Econometrica* 19 (1951): 368–403; Kenneth Arrow and Gerard Debreu, "Existence of an Equilibrium for a Competitive Economy," *Econometrica* 22 (1954): 265–90; Gerard Debreu, *Theory of Value* (New York: John Wiley and Sons, Inc., 1959); F. H. Hahn, *On the Notion of Equilibrium in Economics* (Cambridge: Cambridge University Press, 1973); and H. Scarf, *The Computation of Economic Equilibria* (New Haven, Conn.: Yale University Press, 1973).

2. E. W. Handler, "The Logical Structure of Modern Neoclassical Static Microeconomic Equilibrium Theory." *Erkenntnis* 15 (1980): 33–53.

3. D. W. Hands, "The Logical Reconstruction of Pure Exchange Economics: Another Alternative." *Theory and Decision* 19 (1985): 259–78, and "The Structuralist View of Economic Theories: A Review Essay." *Economics and Philosophy* 1 (1985) 303–35.

4. Kenneth Arrow and Frank Hahn, *General Competitive Analysis* (San Francisco: Holden-Day, Inc., 1971).

5. Gerard Debreu, "Economic Theory in the Mathematical Mode," *American Economic Review* 74, no. 3 (June 1984): 275.

6. Ibid. My italics.

7. Hahn, *On the Notion of Equilibrium in Economics*, p. 10.

8. Nicholas Kaldor, "The Irrelevance of Equilibrium Economics," *Economic Journal*, December 1972, pp. 1241–42.

9. Hahn, *On the Notion of Equilibrium in Economics*, p. 13.

10. Ibid., pp. 14–16.

11. F. H. Hahn, "The Winter of Our Discontent," *Economica*, August 1973, p. 329.

12. Ibid., pp. 40–41.

13. Alan Coddington, "The Rationale of Equilibrium Theory," *Economic Inquiry* 13 (1975): 540–41.

14. Ibid., p. 557.

15. Ibid.

16. Ibid., p. 546.

17. See Imre Lakatos, "The Methodology of Scientific Research Programs," in *Criticism and the Growth of Knowledge* (Cambridge: Cambridge University Press, 1970), pp. 173–77.

18. Ibid.

19. Blaug, *The Methodology of Economics*, p. 192.

20. E. W. Handler, "The Logical Structure of Modern Neoclassical Static Microeconomic Equilibrium Theory," *Erkenntnis* 15 (1974): 33–53.

21. See B. J. Loasby, *Choice Complexity and Ignorance* (Cambridge: Cambridge University Press, 1976), pp. 44–50; also, Daniel M. Hausman, "Are General

Equilibrium Theories Explanatory," in *Philosophy in Economics*, ed. J. C. Pitt (Boston: D. Reidel Publishing Company, 1981), pp. 17–32.

22. See Alexander Rosenberg, "Lakatosian Consolations for Economics," *Economics and Philosophy* 2 (1986): 136.

23. Daniel Hausman, "Are General Equilibrium Theories Explanatory?," p. 21.

24. Weintraub, *General Equilibrium Analysis*.

25. Ibid., pp. 109–10.

26. Ibid., pp. 121–22.

27. But occasionally a promising analysis appears, as in the case of Rudolph Kotter's "General Equilibrium Theory—An Empirical Theory," in *Philosophy of Economics* (Heidelberg: Springer-Verlag, 1982), pp. 103–17. Kotter, acknowledging the deficiencies of general equilibrium theory as a science, argues that the theory is better understood as describing an institutional framework grounded in a social context with its norms and rules. As he puts it: "These norms and rules determine the connection between aims and means in a social context and so they represent social rationality. It is important to maintain that we do not get these norms and rules as statistically assured regularities of behavior, that we do not obtain them by observing what persons are actually doing in certain situations" (pp. 114–15).

28. See Kuhn, *The Structure of Scientific Revolutions*, for much elaboration on this point.

29. Ibid.

Chapter 7. The Postulate of Rationality and Neoclassical Economic Theory

1. See Joseph J. Kocklemans, "Sociology and the Problem of Rationality," in *Rationality Today*, ed. Theodore F. Geraets (Ottawa: University of Ottawa Press, 1979), p. 88. No doubt, the neoclassical economist's appeal to rationality is much more constrained than its idealist definition, according to von Mises, who would argue that all choice making, since it leads to some end, is *ipso facto* rational. See Ludwig von Mises, *Epistemological Problems of Economics* (Princeton: Van Nostrand, 1960).

2. Hempel, *Aspects of Scientific Explanation*, p. 469.

3. Ibid., p. 471.

4. J. Henderson and R. Quandt, *Microeconomic Theory* (New York: McGraw-Hill, 1980), p. 5. Consider, too, C. A. Tisdell's observation that the

> rationality of economic man is fundamental to a substantial body of economic theory. This is so whether we consider neoclassical economic theory as so expertly outlined by Sir John Hicks in *Value and Capital* or more recent developments such as those begun by John von Neumann and Oskar Morgenstern with their publication of the *Theory of Games and Economic Behavior*. Indeed, there is hardly any area of economics in which the rationality postulate is unimportant.

This is found in C. A. Tisdell, "Rational Behavior as a Basis for Economic Theories," in *Rationality and the Social Sciences*, ed. S. I. Benn and G. W. Mortimore (London: Routledge and Kegan Paul, 1976), p. 196. See also

"Comments: Behavioral versus Rational Economics: What You See Is What You Conquer," in *Rational Choice*, ed. Robin M. Hogarth and Melvin W. Reder (Chicago: University of Chicago Press, 1987), pp. 257–65. This interesting essay points out clearly the key issue involved in the analysis of rationality, i.e., the conflict between those who view rationality as a purely formal concept and those who seek to define it in terms of how individuals actually behave.

   5. Alexander Rosenberg, "If Economics Isn't Science, What Is It?," pp. 296–97. Consider Rosenberg's observations:

> It is all well and good to say that economics is conceptually coherent, and that there are no uncontroversial standards against which economics may be found wanting, but this attitude will not make the serious anomalies and puzzles about economic theory go away. These puzzles surround its thoroughgoing predictive weakness.

See, too, Blaug's *Methodology of Economics*, especially pp. 253–64; Benjamin Ward, *What's Wrong with Economics* (London: Macmillan, 1972); Wassily Leontief, "Theoretical Assumptions and Nonobserved Facts," *American Economic Review* 61 (March 1971): 1–7; Alfred Eichner, *Why Economics Is Not Yet a Science* (Armonk, N.Y.: M. E. Sharpe, Inc., 1983); Daniel Bell and Irving Kristol, *The Crisis in Economic Theory* (New York: Basic Books, 1981); Vernon L. Smith, "Theory, Experiment and Economics," *Journal of Economic Perspectives* 3, no. 1 (Winter 1989): 151–69. Consider Smith's observations on the experimental nature of economics:

> Economics as currently learned and taught in graduate school and practiced afterward is more theory-intensive and less observation-intensive than perhaps any other science. I think the statement that "no mere fact ever was a match in economics for a consistent theory" accurately describes the prevailing attitude in the profession (Milgrow and Roberts, 1987, p. 185). This is because the training of economists conditions us to think of economics as an *a priori* science, and not as an observational science in which the interplay between theory and observation is paramount. Consequently, we come to believe that economic problems can be understood fully just by thinking them. (pp. 151–52)

   6. Alexander Rosenberg, *Microeconomic Laws* (Pittsburg: University of Pittsburg Press, 1976), pp. 119–20.
   7. Ibid., p. 124
   8. Ibid., p. 120.
   9. W. V. Quine, "Two Dogmas of Empiricism," *Philosophical Reivew* 60 (1951): 20–43.
   10. Patrick Suppes, "Decision Theory,"*Encyclopedia of Philosophy*, ed. Paul Edwards (New York: Collier Macmillan, 1967), Vol. 2, p. 310.
   11. Amartya Sen, "Rational Fools: A Critique of the Behavioral Foundations of Economic Theory," in *Philosophy and Economic Theory*, ed. Frank Hahn annd Martin Hollis (Oxford: Oxford University Press, 1979), p. 94.
   12. Fritz Machlup, *Methodology of Economics and Other Sciences* (New York: Academic Press, 1978), p. 498.
   13. Ibid., p. 298.
   14. Ibid., p. 281.
   15. Machlup writes: "Examples of *fundamental assumptions* or 'highlevel generalizations' in economic theory are that people act rationally, try to make the most of their opportunities, and able to arrange their preferences in a consistent order; that entrepreneurs prefer more profit to less profit with equal risk" (Ibid., p. 146).

16. Ibid., p. 147.
17. Ibid., p. 145.
18. See Machlup's essay "Positive and Normative Economics," in *Methodology of Economics*, pp. 425-39.
19. Ibid., pp. 438-39.
20. Bruce Caldwell, *Beyond Positivism: Economic Methodology in the Twentieth Century* (London: George Allen and Unwin, 1982), p. 165.
21. Daniel Kahneman and Amos Twersky, "Prospect Theory: An Analysis of Decision under Risk," *Econometrica*, 47 (1979): 263-91.
22. Coldwell, *Beyond Positivism*, p. 167.
23. Frederick Schick, "Rationality—A Third Dimension," *Economics and Philosophy* 3 (1987): p. 54.
24. Ibid., p. 61. But consider the problems generated by this approach: [On the usual analysis, the rationality of choices depends on the utilities and probabilities assigned (and on the agent's beliefs). Some authors are uneasy with this. They feel that a choice based on hastily fixed or perverse or manic utilities or parobabilities deserves no credit. Such a choice cannot be rational. A rational choice, in their broader sense, is a choice that maximizes expected utilities based just on appropriate utilities and probabilities (see, for instance, Elster, 1983, pp. 15ff). What are the criteria of appropriate utilities? What are those of appropriate probabilities? These are dark and difficult questions] ibid., p. 64).
25. Tisdell, "Rational Behavior as a Basis for Economic Theories," p. 219.
26. Hollis, *Philosophy and Economic Theory* p. 13. See, too, Blaug, *The Methodology of Economics*, p. 185.
27. See Herbert Simon, *Models of Bounded Rationality: Behavioral Economics and Business Organizations* (Cambridge: M.I.T. Press, 1982), and "From Substantive to Procedural Rationality," in *Method and Appraisal in Economics*, ed. Latsis, pp. 129-48. It is useful to point out the theoretical exploration of the idea of the limited and constrained rationality of actual agents by Christopher Cherniak in his *Minimal Rationality* (Cambridge: M.I.T. Press, 1986); Cherniak's thesis is to plead the case for a "more realistic model of minimal rationality, where the agent's ability to choose action falls between randomness and perfection" (Cherniak, *Minimal Rationality*, p. 18.)
28. Simon, *Models of Bounded Rationality*, p. 133.
29. Ibid., p. 133.
30. Ibid.
31. Ibid.
32. Tisdell, "Rational Behavior as a Basis for Economic Theories," pp. 196-222.
33. Gary Becker, "Irrational Behavior and Economic Theory," in *The Economic Approach to Human Behavior* (Chicago: University of Chicago Press, 1976), p. 156.
34. T. W. Hutchison, *The Significance and Basic Postulates of Economic Theory* (New York: Augustus Kelley, 1960), p. 42.
35. Blaug, *The Methodology of Economics*, p. 66.
36. Hempel, *Aspects of Scientific Explanation*, p. 167. Hempel's point may be compared with that of Daniel Hausman, who defends the usage of *ceteris paribus* clauses in neoclassical theory by first stating that "generalizations in science are so qualified." It is indeed the case that *ceteris paribus* clauses facilitate theoretical understanding in neoclassical theory, but that is all. The fact is that they serve as an embarrassing reminder to the theorist of the possible differences between the artificial neoclassical world and the real world. Whether *ceteris paribus* clauses are lawlike in structure or otherwise is besides the point. As long as it is recognized that

one of the key assumptions in *cetera* is that "agents are rational" then any claim about the *scientific* usefulness of *ceteris paribus* with regard to questions of causality and the like are rendered suspect. See Daniel Hausman, "Ceteris Paribus Clauses and Causality in Economics," *Philosophy of Science Association Proceedings* (East Lansing, Mich: Philosophy of Science Association, 1989), 308–14.

37. Blaug, *The Methodology of Economics*, p. 95.
38. Ibid., p. 256.
39. Leontief, "Foreword," *Why Economics Is Not Yet a Science*, ed. Eichner, pp. vii–xi.
40. Ibid., p. x.
41. Note the trenchant contemporary critique of instrumentalism by Harvey Leibenstein. According to Leibenstein, instrumentalists are in error to argue "that prediction is the only criterion of really meaningful 'scientific' knowledge." Leibenstein argues rather that

> Predictive capacity without explanatory capacity is worthless. Mere clairvoyance, irrespective of its sharpness, does not iself have scientific standing. Only predictive capacity that arises out of having coherent and communicable explanations has scientific standing. The power to predict is subsidiary to the power to explain. Explanation without prediction is sufficient, but prediction without explanation is of no consequence from a scientific standpoint.

See Harvey Leibenstein, *Beyond Economic Man* (Cambridge: Harvard University Press, 1976), pp. 12–13.

42. Henderson and Quandt, *Microeconomic Theory*, p. 53. Note that "previous analysis" in the above citation refers to orthodox deterministic neoclassical consumer theory.
43. Ibid., p. 54.
44. Consider as a final note on the role of the concept of rationality in the formulation of expected utility decision, theorist Nicholas Rescher's misgivings on the applicability of this approach to human decision making. See Nicholas Rescher, *Rationality* (Oxford: Clarendon Press, 1988), pp. 115–18.
45. Hollis and Nell, *Rational Economic Man*.
46. Amitai Etzioni, *The Moral Dimension—Toward a New Economics* (New York: The Free Press, 1988).
47. Hollis and Nell, *Rational Economic Man*, p. 55.
48. Ibid., p. 60.
49. But note Etzioni's paradigmatic commitment to the idea of a social science expressed in his paper "Toward Deontological Social Sciences," *Philosophy of the Social Sciences* 19 (1989): 145–56.
50. Alexander Rosenberg, "Are Generic Predictions Enough," *Erkenntnis* 30, nos. 1–2 (March 1989): 52–55.
51. Philippe Mongin, "Le réalisme des hypothèses et la *Partial Interpretation View*," *Philosophy of the Social Sciences* 18 (1988): 281–325.
52. Consider the thesis of Hal Varian and Alan Gibbard that models with false assumptions may be vindicated not only when they serve as approximations of reality but as caricatures, when they are structured to highlight or emphasize certain aspects of economic behavior. But again, this approach would be scientifically acceptable only if models as caricatures served the purpose of formulating accurate predictions and satisfactory explanations of economic behavior. See Hal Varian and Alan Gibbard, "Economic Models," *Journal of Philosophy* 75 (1978): 664–77.

Chapter 8. "Positive" Neoclassical Welfare Economics

1. This view, it should be noted, is not held by the majority of economists who still maintain that there is an evident epistemological divide between positive economic theory and welfare economics. But consider the following minority views: G. C. Archibald's "Welfare Economics, Ethics, and Essentialism," *Economica* 26 (November 1959); E. van den Haag's "Normative and Analytical Welfare Economics: Arrow's Pareto Principle and Essentialism," in *Human Values and Economic Policy*, ed. Sidney Hook (New York: New York University Press, 1967), pp. 181–92; and J. R. Hicks, "The Scope and Status of Welfare Economics," *Oxford Economic Papers*, November 1975, pp. 307–26.
2. See Gunnar Myrdal, "What Is Political Economy?," in *Value Judgment and Income Distribution*, ed. Robert Solo and Charles Anderson (New York: Praeger Publishers, 1981), pp. 41–42.
3. A textbook example suffices to demonstrate this point. See James Quirk and Rubin Saposnik, *Introduction to General Equilibrium Theory and Welfare Economics* (New York: McGraw Hill Book Company, 1968), p. 103.
4. Yew-Kwang Ng, *Welfare Economics* (London: Macmillan, 1979), p. 6.
5. Archibald, "Welfare Economics, Ethics, and Essentialism," pp. 316–27.
6. P. Hennipman, "Pareto Optimality: Value Judgment or Analytical Tool," in *Relevance and Precision: From Quantitative Analysis to Economic Policy*, ed. J. S. Cramer, A. Heertge, and P. Venekamp (Amsterdam: North Holland, 1976), pp. 39–69.
7. Ibid., p. 63.
8. J. R. Hicks, "Foundations of Welfare Economics," *Economic Journal*, 49 (1939): 549–52; N. Kaldor, "Welfare Propositions in Economics," *Economic Journal* 49 (1939): 549–52.
9. T. Scitovsky, "A Note on Welfare Propositions in Economics," *Review of Economic Studies* 9 (1941): 77–81.
10. Blaug, *The Methodology of Economics*, p. 146.
11. The term "basic theorem" of welfare economics is that of Amartya Sen's with reference to the fundamental theorem of welfare economic. According to Sen, the basic theorem of welfare economics comprises the direct theorem and the converse theorem. See A. Sen, "The Moral Standing of the Market," in *Ethics and Economics*, ed. by E. Paul et al. (Oxford: Basil Blackwell, 1985), p. 9. Note that instead of the two component theorems of the basic theorem of welfare economics, other authors speak of the two "fundamental" theorems of welfare. See A. Gibbard, "What's Morally Special about Free Exchange," in *Ethics and Economics*, ed. Paul et al., p. 26.
12. See John Rawls, *A Theory of Justice* (Cambridge: Harvard University Press, 1971).
13. Robert Nozick, *Anarchy, State and Utopia* (New York: Basic Books, 1974).
14. See Amartya Sen, *Collective Choice and Social Welfare* (Edinburgh: Oliver and Boyd, 1974).

Chapter 9. Alternative Methodologies

1. See, for example, Ludwig von Mises, *Human Action—A Treatise on Economics* (New Haven: Yale University Press, 1963).

2. James Buchanan, "The Domain of Subjective Economics," in *Method, Process, and Austrian Economics*, ed. Israel Kirzner (Lexington, Mass.: D. C. Heath and Company, 1982), pp. 16–17.

3. Ibid., p. 17.

4. I. M. Kirzner, *Competition and Entrepreneurship* (Chicago: University of Chicago Press, 1973).

5. See Gunnar Myrdal, "The Meaning and Validity of Institutional Economics," in *Economics in Institutional Perspective*, ed. Rolf Steppacher et al. (Lexington, Mass.: D. C. Heath and Company, 1977), pp. 3–6.

6. See, for example, Charles Wilber and Robert Harrison, "The Methodological Basis of Institutional Economics: Pattern Model, Storytelling, and Holism," *Journal of Economic Issues* 13 (December 1979): 899–909.

7. Bruce J. Caldwell, *Beyond Positivism: Economic Methodology in the Twentieth Century* (London: George Allen and Unwin, 1982), p. 204.

8. See James M. Buchanan and Gordon Tullock, *The Calculus of Consent* (Ann Arbor: University of Michigan Press, 1962), for a foundational statement of the public choice school. See, also, the journal *Public Choice* for the ongoing research effort of public choice theorists.

9. See Anthony Downs, *An Economic Theory of Democracy* (New York: Harper and Row, 1957).

10. See Dennis C. Mueller, "Public Choice: A Survey," in *The Theory of Public Choice—II*, ed. James Buchanan and Robert Tollison (Ann Arbor: University of Michigan Press, 1984), pp. 52–55.

11. Note that a substantial amount of the ongoing research efforts of the post-Keynesians is to be found in the *Journal of Post Keynesian Economics*.

12. See J. Kornai, *Economics of Shortage* (Amsterdam: North Holland, 1980), for a useful discussion of the problem of shortages in socialist economies.

13. See Karl Marx and Friedrich Engels, *The Communist Manifesto* (London: Martin Lawrence Limited, 1930), p. 52.

14. See Wlodmierz Brus, "Enterprise and Socialism—Are They Compatible? Lessons from Eastern Europe and China," *Praxis International* 8 (April 1988): 99–108. See, too, Alex Nove, *The Economics of Feasible Socialism* (London: Allen and Unwin, 1983), pp. 118–53, for an earlier exploration of this idea.

Chapter 10. A Theory of Optimal Decision Making

1. But consider Amartya Sen's obervation that "A state can be Pareto optimal with some people in extreme misery and others rolling in luxury, so long as the miserable cannot be made better off without cutting into the luxury of the rich." See Sen, *On Ethics and Economics* (Oxford: Basil Blackwell, 1987), p. 34.

2. See Alan Rowley and Charles Peacock for an instructive analysis of the theoretical distinctions between liberalism and Paretian welfare economics. As the authors put it, "the two philosophies are based upon quite different value judgments" (*Welfare Economics* [New York: John Wiley and Sons, 1975], p. 80). See, too, Amartya Sen, "The Impossibility of a Paretian Liberal," *Journal of Political Economy* 72 (1970): 152–57. Sen's proof in this theorem and the resulting responses are rather surprising given that it is intuitively clear that classical liberalism and Pareto procedures are incompatible, if only because classical liberalism deals with the orthodox negative political freedoms while Pareto procedures refer only to the interpersonal trade of economic assets.

3. James Buchanan, "Political Constraints on Contractual Redistribution," *American Economic Review* 64 (May 1974): 153-57.

4. See Sen, *On Ethics and Economics*. Consider the following: "Moral acceptance of rights (especially rights that are valued and supported, and not just respected in the form of constraints) may call for systematic departures from self-interested behaviour. Even a partial and limited move in that direction in actual conduct can shake the behavioural foundations of standard economic theory" (p. 57). Sen also writes:

> It must, of course, be admitted straightaway that moral rights or freedom are not, in fact, concepts for which modern economics has much time. In fact, in economic analysis, rights are seen typically as purely legal entities with instrumental use rather than any intrinsic value. I have already discussed these neglects. However, it is arguable that an adequate formulation of rights and of freedom can make substantial use of consequential reasoning of the type standardly used in economics. (p. 71)

5. See Rowley and Peacock, *Welfare Economics*, pp. 86-88, for a useful discussion of the classical liberties of liberal thought. But note, too, the traditional tension between the so-called negative and positive liberties:

> Thus, early political philosophy developed the concept of *negative liberties* to express the rights of individuals as having property in their persons: freedom of speech, movement, thought, religion were justified as forms of *freedom from outside interference*. Conversely, the *positive liberties* of rights to participation in governance were to be limited to those who had real property to protect.

See Herbert Gintis, "Social Contradictions and the Liberal Theory of Justice," in *New Directions in Economic Justice*, ed. Roger Skurski (Notre Dame, Ind.: University of Notre Dame Press, 1983), p. 96.

6. Rawls, for example, makes this distinction in *Justice as Fairness*. He writes: "As the conditions for civilization improve, the marginal significance for our good of further economic and social advantages diminishes relative to the interests of liberty, which become stronger as the conditions for the exercise of the equal freedoms are more fully realized." See Rawls, *A Theory of Justice*, pp. 542-43.

7. Note the interesting attempt by Rex Martin to imbue Rawls's idea of right to equal liberty with economic content. Martin argues that the market economy tends to produce unfair results, which eventually threaten Rawls's higher-order principles of justice. Consider the following: "Thus, market results would continually have to be adjusted by the redistributive operation of the difference principle; and measures for shoring up fair equality of opportunity and for controlling the degenerative tendencies of the market would have to be put in place and maintained. There would also have to be devices for reducing or counteracting the growth of economic privilege, with its distorting effect on the equal basic liberties" (Rex Martin, *Rawls and Rights* [Lawrence: University of Kansas Press, 1985], p. 161). Kai Nielsen's attempt to substantiate Rawls's "right to equal liberty" with economic considerations is also noteworthy. Consider his rendering of Rawls's first principle of justice. "Each person is to have an equal right to the most extensive total system of equal basic liberties and opportunities (including equal opportunities for meaningful work, for self-determination and political and economic participation) compatible with a similar treatment for all" (Kai Nielsen, *Equality and Liberty* [Totowa, N.J.: Rowman and Allanheld, 1985], p. 289).

8. Rawls, *A Theory of Justice*.

9. Nozick, *Anarchy, State, and Utopia*.

10. James Buchanan, *The Limits of Liberty* (Chicago: University of Chicago Press, 1975).

11. Orthodox analysis usually distinguishes between the negative and positive freedoms, but it would appear that this distinction is artificial. If the idea of a negative liberty entails the noninterference of the individual's activities within certain spheres, then it would appear that the distinction between negative and positive freedoms is purely arbitrary since the "certain spheres" in question are themselves arbitrary. Proof of this is the varied ways in which such freedoms are defined in the liberal democracies and the statist societies. See Rodney Peffer's "A Defense of Rights to Well-Being," *Philosophy and Public Affairs* 8, no. 1 (Fall 1978): 65–87, for a useful argument pointing out the arbitrariness in the definition of the negative and positive freedoms. See also Alan Hamlin, *Ethics, Economics, and the State* (New York: St. Martin's Press, 1986), p. 107.

12. Rawls, *A Theory of Justice*, pp. 542–43.

13. A useful elaboration of the idea of work as a form of self-actualization is that of David Norton. He argues that "for every person there is a meaningful work which affords to that person intrinsic rewards that will not be exchanged for work of any other sort, or for idleness or unproductive self-indulgence. This work is this individual's self-actualization, and it is likewise productive of social utilities." Norton's eudamonistic thesis is that individuals are principally worth-seeking and that such worth is best obtainable from work which actualizes the individual. See David L. Norton, "Good Government, Justice, and Self-Fulfilling Individuality," in *New Directions in Economic Justice*, ed. Skurski, p. 45. But consider, too, J. Philip Wogaman, *The Great Economic Debate* (Philadelphia: The Wesminster Press, 1977). With reference to Marxism's moral appeal, the author writes: "The essence of human nature is to be creative and social. Man is self-creating. By other creative efforts, we actualize what had been only potential before. Put differently, through our work we exercise our human powers, and only in the exercises of those powers are we fully human" (p. 9).

14. See Barry Clark and Herbert Gintis, "Rawlsian Justice and Economic Systems," *Philosophy and Public Affairs* 7, no. 4 (1978): 302–25. With reference to Rawls's theory of justice, they argue that "In a *Theory of Justice*, however, he has analyzed constitutions only in terms of political and civil rights. We believe that a theory of justice must be premised on the rejection of the dichotomy between the political and economic spheres of social life." The solution is to complement the stock of already guaranteed basic liberties with "An economic bill of rights [which] would assure every citizen of access to decent work, an income sufficient to sustain self-respect, and equal rights in economic decision making" (pp. 324–25).

15. Consider the admirable example afforded by the Japanese economy, in which the individual is practically guaranteed the right to employment. During times of recession, workers are rarely furloughed; instead, they are trained and employed in other areas. For example, during the last recession, furloughed Nissan workers were transferred to productive work in a fish hatchery. See Andrew Zimbalist et al., *Comparing Economic Systems* (New York: Harcourt, Brace, Jovanovich, 1989), p. 44.

16. See Nozick's *Anarchy, State, and Utopia* for an exhaustive defense of the idea of the minimal state.

17. See James Buchanan, "The Coase Theorem and the Theory of the State," in *The Theory of Public Choice—II*, ed. Buchanan and Tollison, pp. 159–73.

18. Note that questions of equity and social justice have been traditionally examined by theorists of welfare economics.

Bibliography

Archibald, G. C. "Welfare Economics, Ethics, and Essentialism." *Economica* 26 (November 1959): 316–27.

Arrow, Kenneth. "An Extension of the Basic Theorems of Classical Welfare Economics." In *Proceedings of the Second Berkeley Symposium on Mathematical Statistics and Probability*, edited by J. Neyman, 507–32. Berkeley: University of California Press, 1951.

Arrow, Kenneth, and Gerard Debreu. "Existence of an Equilibrium for a Competitive Economy." *Econometrica* 22 (1954): 265–90.

Baumol, William. "The Neumann-Morgenstern Utility Index—An Ordinalist View." *Journal of Political Economy* 59 (1951): 61–66.

Becker, Gary. "Irrational Behavior and Economic Theory." In *The Economic Approach to Human Behavior*. Chicago: University of Chicago Press, 1976.

Bentham, Jeremy. *Jeremy Bentham's Economic Writings*. 3 vols. Edited by Werner Stark. London: Allen and Unwin, Ltd., 1952.

Blaug, Mark. "Comment on Douglas W. Hands, Karl Popper and Economic Methodology: A New Look." *Economics and Philosophy* 1 (1985): 286–88.

———. *The Methodology of Economics*. Cambridge: Cambridge University Press, 1980.

Boland, Lawrence. "A Critique of Friedman's Critics." *Journal of Economic Literature* 17 (1979): 503–22.

———. "On the Futility of Criticizing the Neoclassical Maximization Hypothesis." *American Economic Review* 71, no. 5 (December 1984): 1031–36.

———. *The Foundations of Economic Method*. London: George Allen and Unwin, 1982.

Bromberger, Sylvain. "Why Questions." In *Readings in the Philosophy of Science*, edited by Baruch Brody, 60–87. Englewood Cliffs: Prentice Hall, 1970.

Brus, Wlodmierz. "Enterprise and Socialism—Are They Compatible? Lessons from Eastern Europe and China." *Praxis International* 8 (1988): 99–108.

Buchanan, James. "Political Constraints on Contractual Redistribution." *American Economic Review* 64 (May 1974): 153–57.

———. "The Coase Theorem and the Theory of the State." In *The Theory of Public Choice—II*, edited by James Buchanan and Robert Tollison, 159–73. Ann Arbor: University of Michigan Press, 1984.

———. "The Domain of Subjective Economics." In *Method, Process, and Austrian Economics*, edited by Israel Kirzner. Lexington, Mass. D.C. Heath and Company, 1982.

———. *The Limits of Liberty*. Chicago: University of Chicago Press, 1975.

Buchanan, James, and Gordon Tullock. *The Calculus of Consent*. Ann Arbor: University of Michigan Press, 1962.

Caldwell, Bruce. *Beyond Positivism: Economic Methodology in the Twentieth Century*. London: George Allen and Unwin, 1982.

Cartwright, Nancy. *How the Law of Physics Lie*. Oxford: Oxford University Press, 1983.

Cherniak, Christopher. *Minimal Rationality*. Cambridge: M.I.T. Press, 1986.

Clark, Barry, and Herbert Gintis. "Rawlsian Justice and Economic Systems." *Philosophy and Public Affairs* 7, no. 4 (1978): 302–25.

Coddington, Alan. "Positive Economics." *Canadian Journal of Economics* 5 (1972): 1–15.

———. "The Rationale of Equilibrium Theory." *Economic Inquiry* 13 (1975): 539–58.

Comte, Auguste. *Philosophie Positive* vol. 1. Paris: Bachelier, 1830.

Debreu, Gerard. "Economic Theory in the Mathematical Mode." *American Economic Review* 74, no. 3 (June 1984): 267–78.

———. *Theory of Value*. New York: John Wiley and Sons, Inc., 1959.

Dopplet, Gerald. "The Philosophical Requirements for an Adequate Conception of Scientific Rationality." *Philosophy of Science* 55 (1988): 104–33.

Downs, Anthony. *An Economic Theory Democracy*. New York: Harper and Row, 1957.

Eichner, Alfred. *Why Economics Is Not Yet a Science*. Armonk, N.Y.: M. E. Sharpe, Inc., 1983.

Etzioni, Amitai. "Toward Deontological Social Sciences." *Philosophy of the Social Sciences* 19 (1989): 145–56.

———. *The Moral Dimension—Toward a New Economics*. New York: The Free Press, 1988.

Feyerabend, Paul. "Consolations for the Specialist." In *Criticism and the Growth of Knowledge*, edited by Imre Lakatos and Alan Musgrave, 197–230. Cambridge: Cambridge University Press, 1970.

———. *Against Method*. London: New Left Books, 1975.

Frazer, William, and Lawrence Boland. "An Essay on the Foundations of Friedman's Methodology." *American Economic Review* 73 (1983): 129–44.

Friedman, Milton. "Methodology of Positive Economics." In *Essays in Positive Economics*. Chicago: University of Chicago Press, 1953.

Gergen, Kenneth J. "Correspondence versus Autonomy in the Language of Understanding Human Action." In *Metatheory in Social Science*, edited by Donald Fiske and Richard Schweder, 447–87. Chicago: University of Chicago Press, 1986.

Gibbard, Alan. "What's Morally Special about Free Exchange." In *Ethics and Economics*, edited by Ellen F. Paul et al., 20–28. Oxford: Basil Blackwell, 1985.

Gintis, Herbert. "Social Contradictions and the Liberal Theory of Justice." In *New Directions in Economic Justice*, edited by Roger Skurski, 90–112. Notre Dame, Ind.: University of Notre Dame Press, 1983.

Gregersen, Frans, and Simo Køppe. "Against Epistemological Relativism." *Studies in History and Philosophy of Science* 19, no. 4 (1988): 447–87.

Grether, David, and Charles Plott. "Economic Theory of Choice and the Preference Reversal Phenomenon." *American Economic Review* 69, no. 4 (1979): 623–38.

Haag, Ernest van den. "Normative and Analytical Welfare Economics: Arrow's Pareto Principle and Essentialism." In *Human Values and Economic Policy*,

edited by Sidney Hook, 181-92. New York: New York University Press, 1967.
Hahn, Frank H. "The Winter of Our Discontent." *Economica* 40 (August 1973): 322-30.
———. *On the Notion of Equilibrium in Economics*. Cambridge: Cambridge University Press, 1973.
Hahn, Frank, and Martin Hollis. *Philosophy and Economic Theory*. Oxford: Oxford University Press, 1979.
Hamlin, Alan. *Ethics, Economics and the State*. New York: St. Martin's Press, 1986.
Handler, E. W. "The Logical Structure of Modern Neoclassical Static Microeconomic Equilibrium Theory." *Erkenntnis* 15 (1980): 33-53.
Hands, Douglas W. "Karl Popper and Economic Methodology." *Economics and Philosophy* 1 (1985): 83-99.
———. "The Logical Reconstruction of Pure Exchange Economics: Another Alternative." *Theory and Decision* 19 (1985): 259-78.
Hausman, Daniel M. "Are General Equilibrium Theories Explanatory." In *Philosophy in Economics*, edited by J. C. Pitt, 17-32. Boston: D. Reidel Publishing Company, 1981.
———. "Ceteris Paribus Clauses and Causality in Economics." In *Philosophy of Science Association Proceedings*, 308-14. East Lansing, Mich.: Philosophy of Science Association, 1989.
Hempel, Carl G., and Paul Oppenheim. "Studies in the Logic of Explanation." *Philosophy of Science* 15 (1948): 135-75.
———. "Studies in the Logic of Explanation." In *Aspects of Scientific Explanation*. New York: The Free Press, 1965.
———. "The Theoretician's Dilemma." In *Aspects of Scientific Explanation*. New York: The Free Press, 1965.
———. "Typological Methods in the Natural and Social Sciences." In *Aspects of Scientific Explanation*. New York: The Free Press, 1965.
———. "Typological Methods in the Social Sciences." In *Philosophy of the Social Sciences*, edited by Maurice Natanson, 210-30. New York: Random House, 1963.
———. *Aspects of Scientific Explanation*. New York: The Free Press, 1970.
Henderson, James, and Richard Quandt. *Microeconomic Theory*. New York: McGraw-Hill, Inc., 1980.
Hennipman, P. "Pareto Optimality: Value Judgment or Analytical Tool." In *Relevance and Precision: From Quantitative Analysis to Economic Policy*, edited by J. S. Cramer, A. Heertge, and P. Venekamp, 39-69. Amsterdam: North Holland, 1976.
Hicks, John R. "Foundations of Welfare Economics." *Economic Journal* 49 (1939): 549-52.
———. "The Scope and Status of Welfare Economics." *Oxford Economic Papers* (November 1975): 696-712.
———. *Value and Capital*. Oxford: Oxford University Press, 1974.
Hollis, Martin, and Edward J. Nell. *Rational Economic Man*. London: Cambridge University Press, 1975.
Hutchison, Terence W. "History and Philosophy of Science and Economics." In *Method and Appraisal in Economics*, edited by Spiro Latsis, 181-205. Cambridge: Cambridge University Press, 1976.

———. "Professor Machlup on Verification in Economics." *Southern Economic Journal* 22 (1956): 476–83.

———. *The Significance and Basic Postulates of Economic Theory*. London: Allen and Unwin, 1938.

Jevons, William S. *The Theory of Political Economy*. New York: Augustus Kelley, 1965.

Kahneman, Daniel, and Amos Tversky. "Prospect Theory: An Analysis of Decision under Risk." *Econometrica* 47 (1979): 263–91.

Kaldor, Nicholas. "The Irrelevance of Equilibrium Economics." *Economic Journal* 82 (1972): 1237–55.

———. "Welfare Propositions in Economics." *Economic Journal* 49 (1939): 549–52.

Katouzian, Homa. *Ideology and Method in Economics*. New York: New York University Press, 1980.

Kirzner, Israel M. *Competition and Entrepreneurship*. Chicago: University of Chicago Press, 1973.

Knorr-Cetina, Karin. *The Manufacture of Knowledge*. New York: Pergamon Press, 1981.

Kocklemans, Joseph J. "Sociology and the Problem of Rationality." In *Rationality Today*, edited by Theodore F. Geraets, 88–115. Ottawa: University of Ottawa Press, 1979.

Koo, A. Y. C. "An Empirical Test of Revealed Preference Theory." *Econometrica* 30 (1962): 646–64.

Koo, A. Y. C. and G. Hasenkamp. "Structure of Revealed Preference: Some Preliminary Evidence." *Journal of Political Economy* 80 (1982): 724–44.

Koopmans, T. *Three Essays on the State of Economic Science*. New York: McGraw-Hill, Inc., 1957.

Kornai, James. *Economics of Shortage*. Amsterdam: North Holland, 1980.

Kotter, Rudolph. "General Equilibrium Theory—An Empirical Theory." In *Philosophy of Economics*. Heidelberg: Springer-Verlag, 1982.

Kuhn, Thomas. *The Structure of Scientific Revolutions*. 2d. Chicago: The University of Chicago Press, 1972.

Lakatos, Imre. "The Methodology of Scientific Research Programs." In *Criticism and the Growth of Knowledge*, edited by Imre Lakatos and Alan Musgrave. 91–195. Cambridge: Cambridge University Press, 1970.

Latsis, Spiro J. "A Research Programme in Economics." In *Method and Appraisal in Economics*, edited by Spiro J. Latsis. Cambridge: Cambridge University Press, 1976.

Laudan, Larry. *Progress and Its Problems*. Berkeley: University of California Press, 1981.

———. *Science and Hypothesis*. Boston: D. Reidel Publishing Company, 1981.

Leibenstein, Harvey. *Beyond Economic Man*. Cambridge: Harvard University Press, 1976.

Leijonhufvud, Axel. "Schools, Revolutions and Research Programmes." In *Method and Appraisal in Economics*, edited by Spiro Latsis, 65–108. Cambridge: Cambridge University press, 1976.

Leontief, Wassily. "Theoretical Assumptions and Nonobserved Facts." *American Economic Review* 61 (1971): 1–7.

Loasby, B. J. *Choice, Complexity and Ignorance.* Cambridge: Cambridge University Press, 1976.

Machina, Mark J. " 'Expected Utility' Analysis without the Independence Axiom." *Econometrica* 50 (1982): 277–323.

Machlup, Fritz, ed. "Terence Hutchison's Reluctant Ultra-Empiricism." In *Methodology of Economics and Other Social Sciences*, 493–503. New York: Academic Press, 1978.

———. "Paul Samuelson on Theory and Realism." In *Methodology of Economics and Other Social Sciences*, 481–84. New York: Academic Press, 1978.

———. "The Problem of Verification in Economics." In *Methodology of Economics and Other Social Sciences*, 137–57. New York: Academic Press, 1978.

———. "Positive and Normative Economics." In *Methodology of Economics and Other Social Sciences*, 425–50. New York: Academic Press, 1978.

Malinvaud, Edmond. *Lectures on Microeconomic Theory.* London: North Holland Publishing Co., 1972.

Manski, Charles F. "Ordinal Utility Models of Decision Making under Uncertainty." *Theory and Decision* 25, no. 1 (1988): 79–104.

Margenau, Henry, "What Is a Theory." In *The Structure of Economic Science*, edited by Sherman Krupp, 25–38. Englewood Cliffs, N.J.: Prentice Hall, 1966.

Marshall, Alfred. *Principles of Economics.* New York: Macmillan, 1961.

Martin, Rex. *Rawls and Rights.* Lawrence: University of Kansas Press, 1985.

Marx, Karl and Friedrich Engels. *The Communist Manifesto.* London: Martin Lawrence Limited, 1930.

McCloskey, Donald. *The Rhetoric of Economics.* Madison: University of Wisconsin Press, 1985.

McRae, Robert. "The Unity of the Sciences: Bacon, Descartes, Leibniz." In *Roots of Scientific Thought*, edited by Philip Weiner and Aaron Noland, 390–411. New York: Basic Books, 1957.

Mill, James, *Selected Economic Writings.* Edited by Donald Winch. Chicago: University of Chicago Press, 1966.

Mill, John Stuart. *On the Logic of the Moral Sciences.* New York: The Bobbs-Merrill Company, Inc., 1965.

Miroski, Philip. "Physics and the Marginalist Revolution." *Cambridge Journal of Economics* 8 (1934): 361–79.

Mises, Ludwig von. *Epistemological Problems of Economics.* Princeton: Van Nostrand, 1960.

———. *Human Action—A Treatise on Economics.* New Haven: Yale University Press, 1963.

Mishan, Ezra J. *Introduction to Normative Economics.* New York: Oxford University Press, 1981.

Mongin, Philippe. "Are All-and-Some Statements Falsifiable after All." *Economics and Philosophy* 2 (1986): 185–95.

———. "Le réalisme des hypothèses et la *Partial Interpretation View.*" *Philosophy of the Social Sciences* 18 (1988): 281–325.

Mueller, Dennis C. "Public Choice: A Survey." In *The Theory of Public Choice—II*, edited by James Buchanan and Robert Tollison, 23–67. Ann Arbor: University of Michigan Press, 1984.

Myrdal, Gunnar. "The Meaning and Validity of Institutional Economics." In *Economics in Institutional Perspective*, edited by Rolf Steppacher et al., 3–10. Lexington, Mass.: D.C. Heath and Company, 1977.

———. "What Is Political Economy?" In *Value Judgment and Income Distribution*, edited by Robert Solo and Charles Anderson, 41–53. New York: Praeger Publishers, 1981.

———. *Objectivity in Social Research*. London: Gerald Duckworth, 1970.

Nagel, Ernest. "Assumptions in Economic Theory." *American Economic Association Papers and Proceedings* 53 (1963): 211–19.

———. "The Subjective Nature of Social Subject Matter." In *Readings in the Philosophy of the Social Sciences*, edited by May Brodbeck, 39–94. New York: Macmillan, 1968.

———. "The Value-Oriented Bias of Social Inquiry." In *Readings in the Philosophy of the Social Sciences*, edited by May Brodbeck, 98–113. New York: Macmillan, 1968.

Neumann, John von, and Oskar Morgenstern. *Theory of Games and Economic Behavior*. New York: John Wiley and Sons, Inc., 1967.

Newton, Isaac. *Principia Mathematica*. Edited by Florian Cajori. Berkeley: University of California Press, 1960.

Ng, Yew-Kwang. *Welfare Economics*. London: Macmillan, 1979.

Nielsen, Kai. *Equality and Liberty*. Totowa, N.J.: Rowman and Allanheld, 1985.

Norton, David L. "Good Government, Justice, and Self-Fulfilling Individuality." In *New Directions in Economic Justice*. Notre Dame, Ind.: University of Notre Dame Press, 1983.

Nove, Alec. *The Economics of Feasible Socialism*. London: Allen and Unwin, 1983.

Nozick, Robert. *Anarchy, State and Utopia*. New York: Basic Books, 1974.

O'Sullivan, Patrick J. *Economic Methodology and Freedom to Choose*. London: Allen and Unwin, 1987.

Pareto, Vilfredo. *Manual of Political Economy*. New York: Augustus Kelley, 1971.

Peffer, Rodney. "A Defense of Rights to Well-Being." *Philosophy and Public Affairs* 8, no. 1 (Fall 1978): 65–87.

Popper, Karl. *Popper Selections*. Edited by David Miller. Princeton: Princeton University Press, 1985.

———. *The Poverty of Historicism*. New York: Harper Torchbooks, 1964.

Quine, Willard V. "Two Dogmas of Empiricism." *Philosophical Review* 60 (1951): 20–43.

Quirk, James, and Rubin Saposnik. *Introduction to General Equilibrium Theory and Welfare Economics*. New York: McGraw Hill Book Co., 1968.

Rawls, John. *A Theory of Justice*. Cambridge: Harvard University Press, 1971.

Rescher, Nicholas. *Rationality*. Oxford: Clarendon Press, 1988.

Robbins, Lionel. *The Nature and Significance of Economic Science*. London: Macmillan and Co., Ltd., 1932.

Rosenberg, Alexander. "Are Generic Predictions Enough." *Erkenntnis* 30, nos. 1–2 (March 1989): 43–68.

———. "If Economics Isn't Science, What Is It?" *Philosophical Forum* 14 (1983): 296–314.

———. "Lakatosian Consolations for Economics." *Economics and Philosophy* 2 (1986): 127–39.

———. "A Skeptical History of Microeconomic Theory." *Theory and Decision* 12 (1980): 79–93.

———. *Microeconomic Laws*. Pittsburgh: University of Pisttburgh Press, 1976.

Rotwein, Eugene. "On 'The Methodology of Positive Economics.'" *Quarterly Journal of Economics* 73 (1959): 493–613.

Rowley, Alan, and Charles Peacock. *Welfare Economics*. New York: John Wiley and Sons, Inc., 1975.

Russell, Robert, and Maurice Wilkinson. *Microeconomics: A Synthesis of Modern and Neoclassical Theory*. New York: John Wiley and Sons, Inc., 1979.

Salmon, Wesley. *Scientific Explanation and the Causal Structure of the World*. Princeton: Princeton University Press, 1984.

Samuelson, Paul A. "The Empirical Implication of Utility Analysis." In *The Collected Scientific Papers of Paul A. Samulson*, vol. 1, edited by Joseph Stiglitz, 21–34. Cambridge: M.I.T. Press, 1966.

———. "Problems of Methodology—A Discussion." *American Economic Review (Proceedings)* 53 (1963): 231–36.

———. "Professor Samuelson on Theory and Realism: Reply." In *The Collected Scientific Papers of Paul Samuelson*, vol. 3, edited by Robert C. Merton, 765–73. Cambridge: M.I.T. Press, 1972.

———. "Theory and Realism: A Reply." In *The Collected Scientific Papers of Paul Samuelson*, vol. 3, edited by Robert C. Merton, 761–64, 1972.

———. "Theory and Realism: A Reply." *American Economic Review* 54 (1964): 738.

Scarf, Herbert. *The Computation of Economic Equilibria*. New Haven: Yale University Press, 1973.

Schick, Frederic. "Rationality—A Third Dimension." *Economics and Philosophy* 3 (1987): 49–66.

Schoemaker, Paul J. H. "The Expected Utility Model: Its Variants, Purposes, Evidence and Limitations." *Journal of Economic Literature* 20: 529–63.

Scitovsky, Tibor. "A Note on Welfare Propositions in Economics." *Review of Economic Studies* 9 (1941): 77–88.

Scriven, Michael. "The Temporal Asymmetry of Explanations and Predictions." In *Philosopohy of Science*, The Delaware Seminar, vol. 1, 1961–62. Edited by B. Baumrin, 97–105. New York: John Wiley and Sons, Inc., 1963.

Sen, Amartya. "Rational Fools: A Critique of the Behavioral Foundations of Economic Theory." *Philosophy and Economic Theory*, edited by Frank Hahn and Martin Hollis, 86–109. Oxford: Oxford University Press, 1979.

———. "The Impossibility of a Paretian Liberal." *Journal of Political Economy* 72 (1970): 152–57.

———. "The Moral Standing of the Market." In *Ethics and Economics*, edited by Ellen F. Paul et al., 1–19. Oxford: Basil Blackwell, 1985.

———. *Collective Choice and Social Welfare*. Edinburgh: Oliver and Boyd, 1974.

———. *On Ethics and Economics*. Oxford: Basil Blackwell, 1987.

Shafter, Wayne J. "The Nontransitive Consumer," *Econometrica* 42, no. 5 (1974): 913–19.

Simon, Herbert. "From Substantive to Procedural Rationality." In *Method and Appraisal in Economics*, edited by Spiro Latsis, 129–48. Cambridge: Cambridge University Press, 1976.

———. *Models of Bounded Rationality: Behavior Economics and Business Organizations*. Cambridge: M.I.T. Press, 1982.

Smith, Vernon L. "Theory, Experiment and Economics." *Journal of Economic Perspectives* 3, no. 1 (Winter 1989): 151–69.

Suppes, Patrick. "Decision Theory." *Encyclopedia of Philosophy*, vol. 2, edited by Paul Edwards, 310–14. New York: Collier MacMillan, 1967.

Tisdell, C. A. "Rational Behavior as a Basis for Economic Theories." In *Rationality and the Social Sciences*, edited by S. I. Benn and G. W. Mortimore, 196–222. London: Routledge and Kegan Paul, 1976.

Tversky, Amos, and Daniel Kahneman. "Prospect Theory: An Analysis of Decision under Risk." *Econometrica* 47 (1979): 263–91.

———. "The Framing of Decisions and the Psychology of Choice." *Science* 211 (1981): 453–58.

Varian, Hal, and Alan Gibbard. "Economic Models." *Journal of Philosophy* 75 (1978): 664–77.

Wald, Abraham. "On Some Systems of Equations of Mathematical Economics." *Econometrica* 19 (1951): 368–403.

Walras, Leon. *Elements of Pure Economics*. Translated by William Jaffe. Homewood, Ill.: Irwin, Inc., 1954.

Walsh, Vivian. *Introduction to Contemporary Economics*. New York: McGraw-Hill, Inc., 1970.

Ward, Benjamin. *What's Wrong with Economics*. London. Macmillan, 1972.

Weber, Max. "The Interpretive Understanding of Social Action." In *Readings in the Philosophy of the Social Sciences*, edited by May Brodbeck, 19–33. New York: The Macmillan Co., 1968.

Weintraub, E. Roy. *General Equilibrium Analysis: Studies in Appraisal*. New York: Cambridge University Press, 1985.

Wilber, Charles, and Robert Harrison. "The Methodological Basis of Institutional Economics: Pattern Model, Storytelling, and Holism." *Journal of Economic Issues* 13 (December 1979): 1029–37.

Wogaman, J. Philip. *The Great Economic Debate*. Philadelphia: The Wesminster Press, 1977.

Wong, Stanley. "The F-Twist and the Methodology of Paul Samuelson." *American Economic Review* 63 (1973): 312–25.

———. *The Foundations of Paul Samuelson's Revealed Preference Theory*. London: Routledge and Kegan Paul, 1978.

Zeckhauser, Richard. "Comments: Behavior vs. Rational Economics: What You See Is What You Conquer." In *Rational Choice*, edited by Robin M. Hogarth and Melvin W. Reder. Chicago: University of Chicago Press, 1987.

Zimbalist, Andrew, et al. *Comparing Economic Systems*. New York: Harcourt Brace Jovanovich, 1989.

Index

Agent behavior: definitional theorems, 97–98; and ordinal utility theory, 59–60
Agent choice: dimensions in neoclassical model, 61–63, 66, 77, 84; and equilibrium model, 92; formal deterministic theory, 110; model construction for, 145; and personal utility functions, 127; and rationality, 78, 95, 98, 105, 163 n.27; and subjective thoughts, 123
All and some statements, 77, 159 n.45
Allen, R. G. D.: and indifference curve analysis, 51
Analytical propositions, 23; and neoclassical theory, 98–100
Animal behavior: scientific study of 10–11, 152 n.2
Applied science, 17
Archaeology: as applied science, 17, 30
Archibald, G. C.: and welfare economics, 115
Arrow, Kenneth: and equilibrium systems, 87, 91–92; equilibrium theory, 83, 85, 87, 88, 91; freedom and welfare economics, 136; possibility theorem, 116–17
Astronomy: as applied science, 17, 30
Austrian school, 122–24
Axioms of Mental Pathology — A Necessary Ground for All Legislative Arrangements (Bentham), 42

Bacon, Francis: and science of human behavior, 15, 32
Banking system: and revised macroeconomic theory, 147
Becker, Gary: and postulate of rationality, 106

Bentham, Jeremy; economics as science, 39; measurable utilitarianism, 41, 42, 51, 52, 134–35; utilitarian economics, 41–43, 44, 47
Bergson, A.: and freedom and welfare economics, 136
Bernouilli, Daniel: and expected utility theory, 79
Beyond Positivism (Caldwell), 131
Blaug, Mark, 25; and *ceteris paribus proviso*, 108–9; criticism of neoclassical theory, 79, 81; empiricism and economics, 109; and equilibrium theory, 90; ordinal utility theory and revealed preference theory, 58–59, 67; theory falsifiability, 25, 131, 154 n.16; welfare economics as positive economics, 118
Boland, Lawrence: and instrumentalism, 71, 73, 81; and maximization hypothesis, 76–78
Bounded rationality, 105–6
Brain complexity: and decision making, 135–36
Buchanan, James: contractarian, and neoclassical Pareto optimizations, 135; negative liberties, 43; and public choice political economy, 126–27; subjectivity of choice, 123
Budget constraint: and utility theory, 62–63
Bureaucracies: role of state, 130–31

Caldwell, Bruce, 81; empirical analysis and scientific theory, 102; methodological pluralism, 131; and rationality, 101–2, 110
Capital distribution: revised macroeconomic theory, 148
Capitalism: critique of, 128–29

Capital ownership: equalization of, 142
Cardinal utility theory, 42, 47–49, 56, 57–58, 144; critique of, 51–54, 57; and decision making, 103; index, 49–51
Caricatures: as models of prediction, 164 n.52
Cartwright, Nancy: and natural science theories, 30–31
Case studies: and theory testing, 145
"Category mistake," 88
Ceteris paribus proviso, 53, 96–97, 107–9, 150, 151, 163–64 n.36
Cherniak, Christopher: and minimal rationality, 163 n.27
China: and political and economic rights, 141–42
Choice: *a priori* science of, 77, 123; and cardinal utility index, 49–50; consumer and Slutsky equation, 63–64; and equilibrium theory, 84, 86; five axioms of, and expected utility theory, 79; five axioms of, and ordinal utility model, 62–79; and flat indifference curve, 62; and Marxian socialism, 130; normative schedules of, 142; rational, 11, 95–98, 161 n.1, 163 nn. 24 and 27; theoretical approach to uncertainty, 53, 56, 57, 61–62, 157 n.2
Clark, Barry: and Rawlsian justice, 168 n.14
Classical economics, and laws and theories, 41
Classical methodology, 41–44
Coddington, Alan: critique of instrumentalism, 71; equilibrium theory, 88–89, 93
Compensation principle, 118
Competitive behavior, 62
Competitive equilibrium, 84, 85
Comte, Auguste: and social physics study of human behavior, 32
Constraints: and goal programming, 144
Consumer theory: and differential calculus, 57–58; neoclassical, 49–50, 52, 97, 100; and "rational economic man," 60–61, 67; and Slutsky equation, 63–64, 66

Contemporary expected utility theory, 50–51
Contemporary neoclassical economic theory: critical evaluation of, 10; methodological structures, 57–58, 66–67, 79
Contractarians, and optimal decision making, 138–39
Contract theory of liberalism: and neoclassical Pareto welfare economics, 135
"Converse theorem," 119

Das Kapital (Marx), 128
Debreu, Gerard: and equilibrium theory, 83, 85, 87, 88, 91–92
Decision-making theory: and choice rules, 11; and choice schedules, 102–3; evaluative or scientific, 103; microeconomic operational procedures, 144–46; and rationality, 100, 164 n.44; social context of, 135–37; and value assumptions, 11–12
Deductive algorithms: and scientific theory, 27–28
Deductive-nomological (D-N) model of investigation, 16–17, 18, 74–75
Demand functions: and neoclassical theory, 63, 64
Demand model of political decision making, 127
Descartes, René, 14, 15
Descriptivism: and instrumentalism, 68–75
Differential calculus: and consumer behavior theories, 57–58
Dilthey, Wilheim: and phenomenology, 33–34
Diminishing marginal rate of substitution, 52–53
Diminishing marginal utility, 52–53
"Direct theorem," 119
Discourse on Method (Descartes), 14
Distribution: and neoclassical welfare economics, 117–18
Distributive justice: and optimal society, 140–41
Downs, Anthony: and political decision making, 127
Dray, William: and agent choice, 95

INDEX

"Economic man," neoclassical model, 55–56, 101
Economic rights, 141–42, 149
Economics: as policy science, 132; scientific and normative, 105; scientific status of, 25–26, 29, 54
Economic surplus: individual right to, 147, 148, 149
Edgeworth, F. Y.: and indifference curve analysis, 51
Education: compulsory, 139; rights to, 140, 147
Einstein, Albert: and methodology, 16
Elements of Pure Economics (Walras), 46
Empirical knowledge: and technology, 13–14
Empirical proof: and post-Keynesian theory, 128; problems with, 68–70, 76–77, 109, 158 n.26; and science, 15, 16, 20
Employment: rights to, 137–43, 148, 167 n.15
Entelechy, 29
Epistemological relativism, 38, 39, 155 n.19
Epistemology: and natural and social sciences, 37–38, 40
Equality: operational definition of, 141
Equalization of capital ownership, 142
Equilibrium theory: deficiencies as science, 161 n.27; essential features of, 45–46, 84–86; existence of equilibrium, 84–85; as scientific theory, 87–93; stability of, 86; uniqueness of equilibrium, 85
Equity: and neoclassical welfare economics, 117–18
Etzioni, Amitai: and ethics and human economic behavior, 111
Evaluation, social sciences as science, 19–20
Exchange theory, 44–45
Exchange value, 137–38
"Existence of an Equilibrium for a Competitive Economy" (Arrow and Debreu), 85
Existence of equilibrium, 84–85
Expected utility theory, 49–51, 78–79, 110; and Neumann and Morgenstern utility theory, 50, 156 n.19; and rules for rational choice, 11–12
Experimental design: and scientific theory, 29–30
Explanation: and equilibrium theory, 90, 92, 93; logical basis of, 16–18; and market behavior, 106; problems of statistical, 18; and rationality, 104; role in scientific theory, 30, 68–70, 74–75

F-twist, 68–69
Falsifiability criterion, 16, 22, 23, 25, 59, 79, 89, 131, 153–54 n.16
Feyerabend, Paul: and scientific investigation methodology, 18, 19, 20–22, 80
Fiscal policy: and revised macroeconomic theory, 147–48
Fisher, I.: and indifference curve analysis, 51
Formal deterministic agent-choice theory, 110
Formal models: and theory of science, 18
Freedom: and Austrian school, 124; and economic decision making, 143; and Marxian socialism, 130, 131; negative and positive, 168 n.11; restrictions and resistance, 136; and work, 138–40. *See also* Liberty
Friedman, Milton, 131; and instrumentalism and methodology, 67–73

Galbraith, J. K.: and institutionalism, 124
Galileo, 14, 32
Geisteswissenschaften, 9
General equilibrium theory: as a scientific theory, 87–93. *See also* Equilibrium theory
Gibbard, Alan: and models as caricatures and prediction, 164 n.52
Gintis, Herbert: and Rawlsian justice, 168 n.14
Glasnost: and political and economic rights, 141–42, 149
Goal programming: and microeconomic decision-making models, 144
Government: role in economy, 146–47

Hahn, F. H.: and equilibrium theory, 83, 85, 87–89
Handler, E. W.: and equilibrium theory, 84, 90
Hands, Douglas W.: and equilibrium theory, 84
Hausman, Daniel: and *ceteris patribus*, 162, 163–64 n.36; and equilibrium theory, 90–91, 93
Hayek, Friedrich: and Austrian school, 122
Hegel, Georg W. F., 128
Hempel, Carl: and *ceteris paribus proviso*, 108, 163–64 n.36; "ideal types" and scientific research, 35–36; observational-theoretical problem, 154 n.1; rationality and agent choice, 95–96; scientific theory, 27; utility and empirical interpretation, 29
Henderson, James: deterministic agent-choice theory, 110; revealed preference theory and prediction of agent behavior, 58
Hennipman, P.: and welfare economics, 115, 118
Hicks, John R., 131; and compensation principle, 118; marginal rate of substitution, 51, 52–53; static stability, 86
Hobbes, Thomas: social codes, 136; theory of human behavior, 32
Hollis, Martin: and classical-Marxian economic theory and rational economic man, 110–11
"Homo oeconomicus," 43
Houthakker, H.: and ordinal utility theory and revealed preference theory, 58, 69
How the Laws of Physics Lie (Cartwright), 30
Human behavior: nature of, 104; and nonempirical mental states, 31; predictability of, 11; science of, 9, 10, 31–39; subjective dimensions of, 9, 15
Human capital: and economic decision-making, 137–43; investments in, 137–43
Hume, David, 23, 41; empiricism and unity-of-science, 32
Hutchinson, Thomas W.: and *ceteris paribus proviso* in neoclassical theory, 107; economic theory as tautologous, 53–54; scientific status of economic theory, 154 n.21
"Hypothèses non fingo," 14
Hypothesis: "confirmation" or "verification," 59; method of, 15–16; and scientific research, 20; validity of, 67–68
Hypothetico-deductive method, 15

Ideology: and knowledge, 36–37
"Ideal types": and natural and social science research, 35–36
Income: distribution and neoclassical welfare economics, 118; and substitution effects, 64–66
Indifference curve analysis, 51–52; and neoclassical welfare economics, 117
Individual, rights of: and optimal decision-making, 135
Inductive inference: alternative to, 22–24; method of, 15
Inductive-statistical (I-S) model of investigation, 16–18
Institutionalism: alternative to neoclassical theory, 124–25
Instrumentalism: and descriptivism, 67–78; methodology of, 15, 16; and scientific investigation, 15–16, 19, 153 n.9, 164 n.41
Instrumental rationality, 95
Intellectuals: and socialist revolutions, 130
Intensity: preference, 49; ratio of, 46
International factor-price equalization, 72
Interpersonal comparisons of utility, 116–17, 118, 140
Interpretive research schools, 9
Investigation: methodology of, 15, 16; models of, 16–17. *See also* Methodology; Research
Invisible-hand, neoclassical theory, 149
Irrational behavior: and theory construction, 92
"Irrational" decision makers, 106
"Irrefutable hard core," 79–80

Jevons, William S.: economics as empirical science, 39, 114; neoclassical economic revolution, 43, 55, 131; theories of utility and exchange, 44–45, 47, 48

Justice, and liberties, 137, 167 nn. 6 and 7, 168 n.14

Kahneman, Daniel: and expected utility theory, 79; maximization and postulate of rationality, 102
Kaldor, N.: compensation principle, 118; returns to scale and equilibrium theory, 87
Keynes, John Maynard: and labor costs, 148–49
Keynesian theory: and government role in economy, 146, 148
Kirzner, Israel M.: and Austrian school, 124
Knowledge: acquisition and types of, 13–14; role of ideology in, 36–37; scientific, 164 n.41
Koopmans, T., 68
Kuhn, Thomas, 18, 19; preparadigmatic stages of social sciences, 37; protoscientific theories, 39; and scientific investigation methodology, 20–22, 23, 25, 26, 80, 131

Lakatos, Imre, 18, 22; and "research program," 24–25, 26, 89, 91–92; and scientific theory and methodology, 79–81, 131
Latsis, S. J.: and "protective belt" of research program, 25
Laudan, Larry, 15; instrumentalist, 26; scientific theory and methodology, 22, 27
Law of variation of utility, 44
Leibenstein, Harvey: critique of instrumentalism, 164 n.41
Leijonhufvud, Axel: and Keynesian economic theory, 25–26
Liberalism: and welfare economics, 134–35, 166 n.2, 167 n.5
Liberal theorists: and rights, 141
Libertarians: and government role in economy, 146
Liberty: and optimal decision making, 135; and social collectivity, 120–21, 137–40
Limited and constrained rationality, 163 n.27
Little, I. M.: and welfare economics, 136
Locke, John, 41; science of human behavior, 32; social codes, 136

Logical positivists, 23
"Lysenko affair," 37

McCloskey, Donald: and method of communicating, 81
Mach, Ernst: and methodology, 16
Machlup, Fritz: postulate of rationality as heuristic principles, 100–101, 102, 110; rejection of descriptivism, 72–73; validation of economic propositions, 54
McKenzie, L.: and equilibrium theory, 91, 92
Macroeconomic theory, revised, 146–50
Malthus, Thomas R., 128; and economic laws and theories, 41; political economy, 114
Mannheim, Karl: classless intellectuals, 37; ideology and knowledge, 36
Margenau, Henry: and theoretical constructs in natural and social sciences, 54–55
Marginalist revolution: and neoclassical economics, 43, 47
Marginal rate of substitution, 52–53
Marginal utility theory, 44–45, 52, 53 and exchange, 44–45
Market: and rationality, 106
"Marketplace": public-choice theory, 127
Marschak, Jacob: and expected utility theory, 79
Marshall, Alfred: measurement of utility, 47, 48; static stability, 86
Marx, Karl: and capitalism, 128–29; ideology and knowledge, 36; science of human behavior, 32
Marxian socialism: alternative to neoclassical theory, 128–31; and falsification criterion, 23–24; methodology of, 125–26; and rights, 141
Mathematical expression: and microeconomic decision-making models, 144; scientific study, 15, 16
Maximization of utility: concept limitations, 43–44, 46, 52, 76–78, 97, 145; and profits and policy science, 132; and rationality, 95, 100
Maximizing objective functions, 144
Maximum ophelimity, 43
Measurement: ordinal utility theory,

51–52; and scientific theory, 27–28; utility theory, 41, 42, 46–54, 134–35, 156n.19
Menger, Carl: and Austrian school, 122; neoclassical economic revolution, 43, 55; utilitarian theory, 47
Methodology: and contemporary neoclassical economic theory, 57–58, 66–67, 79; empirical problems, 28, 69–70, 88, 154n.1, 158n.26; instrumentalism, 15, 16, 67–73; and Marxian socialism, 130; natural and social sciences, 36, 54–55, 71, 80–82; neoclassical economic, 41–43, 55–56, 79, 156n.5; neoclassical research program, 150; and ordinal theory, 57–60, 67–78, 157n.1; pluralism, 131; political economic orientations, 125–27; pre- and post-Galilean era, 38; realist, 15, 16; retrospective questions, 131–32; revealed preference, 53, 58–60; scientific investigation, 18–19, 20–22, 79–81, 131; standards of, 18–19, 41–43, 55–56; writers of, 16
Methodology of Economics (Blaug), 131
"Methodology of Positive Economics" (Friedman), 67
Microeconomic models, revised, 143–46
Mill, James, 42
Mill, John Stuart: economics as moral science, 114; mental laws, 32, 55; utilitarian economics, 41–42, 44
Minimal rationality, 163n.27
Minimizing objective functions, 144
Mirowski, Philip: and neoclassical theory and Newtonianism, 156n.5
Mises, Ludwig von: and Austrian school, 122; and rational choice making, 161n.1
Mongin, Philippe: and "partial interpretation" view of theory construction, 112
Moral Dimension-Toward a New Economics, The (Etzioni), 110–11
Morgenstern, O.: and cardinal utility index, 49, 50

Myrdal, Gunnar: and institutionalism, 124

Nagel, Ernest: critique of instrumentalism, 70–71; single methodology of investigation, 34–35; social science methodology, 54
Natural sciences: and *ceteris paribus proviso*, 108; and ideology, 37; skepticism of, 30–31; and social sciences, 36–38, 54, 151
Negative heuristic, 79; scientific research program, 24
Negative liberties, 135, 137–38, 139
Nell, Edward J.: and classical-Marxian economic theory and rational economic man, 110–11
Neoclassical economic methodology: additive and cardinal utility, theory, 47–48; diminishing marginal utility, 52–53; marginal utility and exchange, 44–45; marginalist revolution, 44–47; maximum utility, 43; measurement of utility, 47, 50, 51, 54; and Newtonianism, 42–43, 55–56, 156n.5; ordinal utility, and indifference curve analysis, 51–52; principle of diminishing marginal rate of substitution, 52–53; theory of equilibrium, 45–46; theory of revealed preference, 53; theory rejection, 79; utility theory and cardinal utility index, 49–51, 156n.19
Neoclassical economic theory: birth of, 43, 44, 46, 55–56: and *ceteris paribus proviso*, 107–9; cultural limitations, 104; and empiricism and assumptions, 69–70; and employment, 148; and equilibrium, 90–91; and falsificaition criterion, 153–54n.16; "hard core," 25; invisible-hand approach, 149; limitations of, 143; and Marxian political economic theory, 126; and Marxian socialism, 130; and maximization hypothesis, 76–78; and nomological model of scientific explanation, 74–75; and operational structures of, 64–67; and optimal decision making, 142; and

ordinalist-choice, 67–78; and postulate of rationality, 96–107; problems of, 9–10; scientific basis of, 96–98, 102–3, 109–10, 112; Slutsky equation, 64

Neoclassical general equilibrium theory: assumptions of, 84; methodology, 45–46; and post-Keynesian school, 128

Neoclassical microeconomic theory: normative assumptions, 119–20; ordinal utility theory and agent behavior, 59–60; and welfare economics, 134

Neoclassical model: and Marxian political economy, 125–26, 127; and neoclassical political economy, 126–27; and neoclassical political economy, 126–27; and public-choice political economy, 126–27; structural elements of, 62–64

Neoclassical ordinal theory: and neoclassical welfare economics, 117

Neoclassical political economy: methodology of, 125–26; neoclassical Paretian welfare economics, 126

Neoclassical welfare economics: neoclassical political economy, 126; and positive economic theory, 117; structure of, 115–18

Neopositivists and phenomenologists, 31–32

Neovitalism, 29

Neumann, John von: cardinal utility index, 49, 50; equilibrium theory, 85; expected utility theory, 79

Newton, Issac: experimental mechanics, 32; "hypotheses non fingo," 14; and methodology, 16; second law, 99

Newtonianism: and neoclassical economic methodology, 42–43, 55–56, 156 n.5; research standard, 41, 42–43, 55–56

Normal science: and neoclassical economic theory, 109–10

Normative dimensions: and economic decision making, 11–12, 107, 151, 152–53 n.5; economics, 9; and Marxian socialism 129: neoclassical microeconomic theory, 119; and neoclassical welfare economics model, 117, 118, 120–21; and positive economics, 101; and rationality, 100, 105

Normative naturalism, 39, 155 n.21

Norton, David: and work as self-actualization, 168 n.13

Novel theories: role of, 21–22

Nozick, Robert: ethical research, 120–21; liberties and social-contract theories, 136, 137–38

Observation: and scientific theoretical problems 28, 154 n.1

Occam's razor, 68, 103

Optimal economic decision-making theory: parameters of, 143

Optimal firm: and neoclassical economic theory, 64–67

Optimality, Pareto, 116, 166 n.1

Optimal strategies, 144

Optimization of welfare: role of state, 149

Ordinal utility theory, 49, 50, 51–53, 57–60, 144, 150; axiomatic structure of, 60–62, 79; demand function, 63; and indifference curve analysis, 51–52; and methodology, 57–60, 67–78, 157 n.1; optimal firm, 64–67; Slutsky equation and consumer choice, 63–64; and social-welfare function, 116; weakness of, 29

Orthodox socialist economic theory: scientific basis of, 129

Ownership: and Marxian socialism and means of production, 130, 142, 150

Pain and pleasure: quantification of utility, 41, 42, 51, 52

Paradigm: and "theory confirmation," 20–21

Paretian welfare economics: limitations of, 134–35, 166 n.2, 167 n.5; and microeconomic theory, 119–20, 121

Pareto, Vilfredo, 131; and competitive equilibrium 85; economic equilibrium, 43; efficiency, 120; indifference curve analysis, 51; optimal welfare functions, 140–41; optimality, 116, 134, 135; optimality and decision making, 136–37; welfare economics, 114–18

Peacock, Charles: and liberalism and Paretian welfare economics, 166n.2, 167n.5
Perestroika: and political and economic rights, 141–42, 149
Phenomenology: and neopositivists, 31–32; and social sciences, 20, 36
Philosophy of science, postscript, 39–40
Physics: and scientific theory, 31
Platonic theory of forms, 13
Policy: and neoclassical theory, 132
Political economy, 114; alternative to neoclassical theory, 125–27
Political rights: as avenue for economic rights, 141–42
Popper, Karl, 18, 19; falsifiability thesis, 22, 27, 79, 89, 90, 153–54n.16; scientific methodology, 22–24, 25, 26, 34, 59, 79, 80, 131
Positive economic theory: and neoclassical welfare economic theory, 117; normative dimensions, 101, 118
Positive equilibrium, competitive price, 84
Positive heuristic, 79; scientific research program, 24
Positive microeconomics model: and Paretian welfare economics, 119
Positive neoclassical theory: and welfare economics, 117
"Positive welfare economics": contradiction of, 118
Positivism: and economic methodology, 131–32; logical, 23; social-science counterpart of Newtonianism, 43; and social sciences, 20
Possibility theorem, 116–17
Post-Keynsian model: alternative to neoclassical theory, 127–28
Postscholastic research paradigms, 14–15
Postulate of rationality, 150; and neoclassical theory, 96–107, 132; normative dimensions of, 100, 105
Prediction: and Austrian school, 123–24; and economic theory, 112; and equilibrium theory, 90, 92, 93; and human behavior, 31; and institutionalism 124; and Marxian socialism, 125, 129; and model

caricature, 164n.52; and neoclassical theory, 107–10; and scientific theory, 30, 39–40, 67–70, 75–76, 158n.26, 164n.41; and theoretical status, 71–72
Preference theory: and utility measurement, 49, 52, 53, 156n.19
Prescriptivism: and rationality, 94, 99, 100, 104, 105, 108. 110; and scientific theory, 23
Price: and consumer behavior, 66; and equilibrium theory, 85–86
Principia (Locke), 41
Principia (Newton), 14
Principle of maximization, 111
Privatization: effects of, 149–50
"Problems of Methodology—A Discussion" (Samuelson), 67, 72
Procedural rationality, 105–6, 163n.27
Productive labor: as a right, 138–39
Profit: maximization and rationality, 97–98; and optimal firm, 65
"Protective belt," 79–80, 91–92
Protoscientific theories, 38–39
Proudhoun, Pierre Joseph, 128
Psychology: and epistemological relativism, 39, 155n.20
Public-choice political economy: methodology of, 125–27
Public-choice theorists: and government role in economy, 146
Public officials: role in public-choice theory, 126–27
Pure science, 17

Quandt, Richard: deterministic agent-choice theory, 110; revealed preference theory and prediction of agent behavior, 58
Quantitative and qualitative expression, 15, 16
Quantity theory of money in monetary economics, 71–72
Quasi-scientific, 17
Quine, Willard V.: and analytic and synthetic propositions, 99

Rate of product transformation (RPT), 65
Rate of technical substitution (RTS), 65
"Rational action": definition of, 111, 162n.15

INDEX

Rational behavior: and "empathic understanding," 33; "ideal type," and social science investigaton, 34, 35, 36

Rational choice: defined by postulate of rationality, 96–107, 163 nn. 24 and 27

Rational decision making: alternative theories of, 105–6, 164 n.44

Rational economic man, 96, 161 n.4

Rational Economic Man (Hollis and Nell), 110–11

Rationalism 15, 16

Rationality, 94–96; Austrian school, 123; and decision theory, 100; elimination of, 144; as normative proposition, 100; a postscript on, 110–12; substantive, 106, 107; and theory construction, 103, 104

Ratio of intensities, 46

Rawls, John: ethical research, 120–21; liberties and social-contract theories, 136, 137, 138–40; principles of justice, 139, 167 nn. 6 and 7, 168 n.14

Realist methodology, 15, 16

Relativism: post-Kuhnian epistemological, 38, 39; and scientific truth, 20–22

Rescher, Nicholas: and rationality and expected utility decision, 164 n.44

Research: differences between natural and social sciences, 34, 36, 37–38, 54–55, 151

Research paradigms: and empirical sciences, 109

"Research program," 24–25, 26, 89, 91–92, 150

Revealed preference theory: methodology of, 53, 58–60; and pure ordinal theory, 69

Ricardo, David, 128; and economic laws and theories, 41, 55; political economy, 114

Rights: economic content, 149, 167 n.7, 168 n.14; and economic decision making, 137–43; and revised macroeconomic theory, 147; work, 137–43, 148, 167 n.15

Risk: and cardinal utility index, 49–50; and expected utility theory, 78–79

Robbins, Lionel: and measurement of utility, 51

Rosenberg, Alexander: analytical propositions and neoclassical theory, 98, 99; criticizes scientific pretensions of neoclassicism, 81; and equilibrium theory, 90; "generic predictions" of economic theory, 112

Rotwein, Eugene, 68

Rousseau, Jean Jacques, 128

Rowley, Alan: and liberalism and Paretian welfare economics, 166 n.2, 167 n.5

RPT (rate of product transformation), 65

RTS (rate of technical substitution), 65

Russell, Robert: and scientific status of neoclassical theory, 66–67

Salmon, Wesley, 18

Samuelson, Paul, 131; measurement of utitlity, 51; ordinal theory of utility, 52–53; and descriptivism, 68–74; and revealed preference theory, 53, 58, 58–59

Say, Jean Baptiste: and economic laws and the theories, 41

Scarf, Herbert, 83

Schick, Frederick: and relativity of rationality, 102

Science: birth of, 38–39; differences between natural and social, 34, 36, 37–38, 54–55, 151; methodology of, 22–26; and neoclassical economic theory, 109–10; philosophy of, 39–40

Science of economics: normative dimensions of, 11–12

Science of human behavior, 9, 10, 31–39

Scientific change: theories of, 155 n.19

Scientific economics, 9

Scientific investigation: D-N model, 16–17; historical overview, 13–20; I-S model, 16–17; instrumentalist and realist approaches, 15, 16; recent methodologies, 20–22

Scientific theory: and case studies, 145; constitution of, 27–31, 39–40; dimensions of, 68–71, 74, 81–82;

disconfirmation of, 102; general equilibrium theory, 87–93; Marxian political economics, 126; and Marxian socialism, 129, 130; nature of proof, 20–22; requirements of, 103–4, 112, 131–32; verification of, 19

Scitovsky, Tibor: and compensation principle, 118

Second principle of justice, 120

"Second victim," 52

Self-actualization, right to, 138–40, 168 n.13

Sen, Amartya: basic theorem of welfare economics, 166 n.11; ethical research, 120; moral rights, 166 nn. 1 and 2, 167 n.4; problematic scientific status of neoclassical theory, 100

Shafer, Wayne J.: and transitivity assumption, 158 n.7

Simon, Herbert, 68; and bounded and procedural rationality, 105–6, 144

Slutsky equation: and consumer choice, 63–64, 66

Smith, Adam, 128; and economic laws and theories, 41, 55; political economy, 114

Smith, Vernon: and experimental nature of economics, 162 n.5

Social collectivity: contours of, 144; and revised macroeconomic theory, 147

Social context: and human behavior, 32–33

Social-contract theories: and rights, 136–37

Social framework: and economic decision making, 120–21

Socialist economics model: and neoclassical economics program, 130

Social justice: and economic decision making, 137–43

Social sciences: and ideology, 36–37; proliferation of schools of thought, 36, 37; quasi-scientific, 17; scientific methodological criteria of, 19–22, 29; unity-of-science approach, 32, 34–35, 54

Social scientist: dilemma of, 32–33

Social state, Pareto optimal, 134

Soviet Union: and political and economic rights, 141–42

Stability of equilibrium, 86

Standards, methodological, 18–19

State, Pareto optimal, 166 n.1

State planners: and orthodox socialist model of economics, 129

Statistical explanations: problems of, 18

Statistics: and social science investigation, 35

Structure of Scientific Revolutions (Kuhn), 18, 20

Subjectivity: and social science inquiry, 32–35, 54

Substantive rationality, 106, 107

Suppes, Patrick: and decision theory and rationality, 100

Supply models: and level of government, 127

Synthetic propositions, 23

Tâtonnement adjustment mechanism, 86

Theory: confirmation, 20–21, 54, 59, 80; of exchange, 44–45; falsifiability of, 131; of general equilibrium 46; of gravitation, 14; of justice, 139, 167 nn. 6 and 7, 168 n.14; metaphysical basis, 16; observational problems, 28, 154 n.1; of revealed preference, 53; of science, new approaches, 18–19; of scientific change, and epistemological relativism, 155 n.19; scientific dimensions of, 68–69; of utility maximization, 68–69

Theory of Games and Economic Behavior (von Neumann and Morgenstern, 49

Theory of Political Economy (Bentham), 44

Theory of Value (Debreu), 85

Tisdell, Clem: neoclassical rational decision-making theory, 106, 161 n.4; rationality and prediction, 104

Transitivity: and cardinal utility index, 50; and ordinal utility theory, 61, 158 n.7

Treatise of Human Nature (Hume), 23, 32
Tullock, Gordon: and public choice political economy, 126–27
Tversky, Amos: and expected utility theory, 79; maximization and postulate of rationality, 102

Uniqueness of equilibrium, 85
Unity-of-science approach, 32-36
Utilitarian economists, 17, 41–42, 44, 46, 47
Utility theory: and cardinal utility index: 49–51; elimination and substitution of, 144; and expected theory, 50, 110, 156 n.19; and intentions, 102–3; measurement of, 42–54, 56, 134–35, 156 n.19; measurement of, and cardinalist approach, 47–51; rationality, condition for, 97–98, 163 n.24; subjective aspects, 123; and welfare economics, 134–35

Value: bias in social science, 32–35, 100, 132; exchange, 137–38; goods and services, classical economics, 43–44; value judgments: Marxian political economics, 126; neoclassical microeconomic theory and Paretian model, 119–20, 166 n.2
Varian, Hai: and models as caricatures and prediction, 164 n.52
"Veil of ignorance," 120
"Verstehen," 33

Wald, Abraham: and equilibrium theory, 83, 85
Walras, Leon, 83, 131; economics as empirical science, 114; neoclassical economic revolution, 43, 55; static stability, 86; theory of equilibrium, 45–46, 47
Wealth, distribution of: and classical economics, 43
Weber, Max: science of human behavior, 32; and "verstehen," 33
Weintraub, E. Roy: neoclassical general equilibrium theory, 81, 91, 92–93; neoclassical theory, 25–26
Welfare: government role in optimization, 149
Welfare economics: basic theorem of, 165 n.11; limitations of, 134–35, 166 n.2, 167 n.5; question of, 114–15
Wilkinson, Maurice: and scientific status of neoclassical theory, 66–67
Wong, Samuel: and descriptivism and scientific explanations, 74; instrumentalism and realism, 73; ordinal utility theory and revealed preference theory, 58–59; and preference theory, 53
Work: and liberty, 138–39, 147, 151; and self-actualization, 138–40, 168 n.13

Yew-Kwang Ng: and welfare economics, 114–15